THE OPEN UNIVERSITY
Science: A Third Level Course
The Nature of Chemistry

The Structure of Chemistry

Prepared for an Open University Course Team

The Open University Press

The S304 (S351) Course Team

Chairmen

L. J. Haynes (*to January 1975*)
D. A. Johnson (*from January 1975*)

General Editor

Christina Warr

Unit Authors

A. R. Bassindale
S. W. Bennett
C. J. Harding
L. J. Haynes
R. R. Hill

D. A. Johnson
Joan Mason
D. R. Roberts
C. A. Russell (*Faculty of Arts*)

Editor

Eve Braley-Smith

Other Members

K. Bolton
M. J. Bullivant (*Course Assistant*)
R. M. Haines (*Staff Tutor*)
Janet E. Haylett (*Course Assistant*)
D. A. Jackson (*BBC*)
G. W. Loveday (*Staff Tutor*)

L. Melton (*Library*)
Jane Nelson (*Staff Tutor*)
R. C. Russell (*Staff Tutor*)
J. Stevenson (*BBC*)
R. J. K. Taylor
B. G. Whatley (*BBC*)

Consultants

R. M. Evans (*Glaxo Laboratories*)
P. Farago (*The Chemical Society*)
H. Heal (*Queen's University of Belfast*)
D. R. Marshall (*University College of North Wales, Bangor*)
R. Maskill (*University of East Anglia*)

The Open University Press,
Walton Hall, Milton Keynes.

First published 1976.

Copyright © 1976 The Open University.

Designed by the Media Development Group of the Open University.

Printed in Great Britain by
Martin Cadbury, a specialized division of Santype International,
Worcester and London.

ISBN 0 335 04230 9

This text forms part of an Open University course. The complete list of Units in the course is printed at the end of this text.

For general availability of supporting material referred to in this text please write to the Director of Marketing, The Open University, P.O. Box 81, Walton Hall, Milton Keynes, MK7 6AT.

Further information on Open University courses may be obtained from the Admissions Office, The Open University, P.O. Box 48, Walton Hall, Milton Keynes, MK7 6AB.

1.1.

UNITS 1–3 have been prepared on the basis of
unpublished research by COLIN A. RUSSELL

Unit 1 One chemistry—or two?

Contents

Figure 1.1 A. L. Lavoisier (1743–94), one of the founders of modern chemistry. He was one of the discoverers of oxygen, which he recognized as the substance taken up during combustion and respiration. His execution by guillotine at the height of the French Revolution Terror was as much a tragedy for chemistry as it was a travesty of justice.

Figure 1.2 Joseph Priestley (1733–1804), English radical reformer, unitarian minister and pioneer of the chemistry of gases, discovering oxygen by heating mercuric oxide. Like Lavoisier he suffered at the hands of mob violence, though in his case it was his own radicalism that caused the destruction of his house and chapel in Birmingham. His last ten years were spent in America.

1.0 Introduction

1.0.1 Some preliminary questions

As you will now be aware this Course is an enquiry into the nature of chemistry. It is not an arbitrary collection of snippets of odd chemical information reflecting either the interests of the individual authors or the whims of the moment (chemists, like everyone else, suffer from occasional lapses into mere 'trendiness' just for the sake of it). Nor is it a distillation of the very latest pieces of chemical news, the ultimate in modernity; if that were so it would have to be rewritten at least once a year. And you will not find that it is even a propaganda exercise mounted to demonstrate either the relevance of chemistry for society today or the inherent fascination of chemical logic. These two elements are hopefully present but they have not dictated the shape or structure of the Course. Instead everything that is presented to you is there because of its potential contribution to an understanding of the nature of chemistry itself. It is important in its own right, but it must be seen in terms of that ultimate aim.

In a sense, therefore, this whole Course is concerned to pose one question: what is chemistry? The purpose of this block of three Units is to show that, concealed within that simple question, are several others that first need attention. Then, having clarified what exactly it is that we are asking, we shall go on to demonstrate the relevance to our quest of one crucial factor that is surprisingly often neglected. This will enable us to pursue a line of preliminary enquiry so that we shall finally be in a position, if not to restate the question in slightly different terms, at least to rethink the answers within a new perspective. The rest of the Course will then address itself to the question in the light of what we are able to do in these three Units.

It is well to indicate at once that the new perspective for these three Units is in essence a *historical* one. But it is vitally important to stress that the distinction between the subject matter of these three Units and the ones following is one of *degree* rather than *kind*. There is certainly more reference to the past in these Units than elsewhere, but it would be totally misleading to suggest that here we are dealing with history and all the other Units are concerned with chemistry. It is *all* about chemistry. The fallacy in the distinction that we have just rejected is simply this: chemistry, however we define it, is something dynamic and constantly changing. The chemistry of today is not the same as that of yesterday. At this time last year it was quite noticeably different in some major respects. A century ago the differences were enormous; the electron had not yet been discovered and so found no place in chemical theories, atoms were commonly disbelieved in by the inorganic chemists, and the less sceptical organic chemists were having to come to terms with the absurd novelty that *their* carbon atoms had valencies directed towards the corners of a mythical tetrahedron. A hundred years before that the discoveries of Lavoisier and Priestley were just about to topple the whole edifice of chemical thought as a new element, oxygen, emerged from their researches. Go back a century more and chemistry will be scarcely recognizable except to the trained eye of the historian. It is obvious that chemistry has a past, and it can only need a moment's reflection to realize that that past has shaped the present. Usually, because of the enormous pressure of time, chemistry courses are unable to afford anything but the most cursory glance at the earlier stages of chemical development. As a result the experience can be rather like examining the last frame of a ciné film in isolation from all that went before. In this Course we have no intention of playing the whole film for you but we do hope to show sufficient of the past to give you a much more penetrating insight into the nature of chemistry than you would get from a totally non-temporal presentation, whether 'frozen' in 1975, 1976 or at any other recent date. Most, though not all, of this material will occur within these three Units.

In trying to pursue this course of action we shall have to avoid two extreme points of view. On one hand there is the approach of some historians of science which, because of their preoccupation with the past and their concentration on the historical processes, leads them to undervalue the massive achievements of modern science. Understandably dazzled by the complexities of current chemistry they tend to close their eyes to the present—and thereby fail to offer anything of interest, value or relevance to those concerned to understand the state of the subject today. Equally to be avoided is the cavalier attitude of some chemists who, looking for a 'breakthrough' in their own area of research, overlook the temporary nature of all scientific theories and tend to despise the past. A contemptuous, or even a patronizing, attitude to workers of earlier times, regarding them as in some way deficient for not having grasped 'the truth', is precisely the right condition of self-blindness required to militate effectively against any real understanding of the nature of chemistry today.

> The object of this work is to exhibit as complete a view as possible of the present state of chemistry; and to trace at the same time its gradual progress from its first rude dawnings as a science, to the improved state which it has now attained. By thus blending the history with the science, the facts will be more easily remembered, as well as better understood; and we shall at the same time pay that tribute of respect to which the illustrious improvers of it are justly entitled.

These words, from Thomas Thomson's *System of Chemistry* of 1820, express a more ambitious design than we have in the Units, but they do catch the spirit of our approach. We shall try to blend 'the history with the science'.

With these preliminaries over we can now address ourselves to the question *what is chemistry*? We could, of course, consult a dictionary, and some dictionary definitions appear below. But it is unlikely that this would be a good way for capturing the essence and spirit of something as rapidly changing as chemistry. Nor do we get much help from the derivation of the word. It comes from the term given to its predecessor, *alchemy*, and this in turn may have been derived from the Greek work *chemeia*, 'black', referring to the black soil bordering the River Nile, and thus a Greek term for Egypt, whence alchemy was supposed to have come. But this is far from certain and recent sponsors have been found for the counter-suggestion that alchemy was derived from a term used in ancient China! In any case etymology is often a poor guide to current practice.

Figure 1.3 Thomas Thomson (1773–1852), Scottish chemist and founder of an important research school at Glasgow. He is chiefly famed for his early popularization of Dalton's atomic theory.

> To show that this question raises all sorts of subsidiary issues, here is a list of some suggested answers, not all as frivolous as they might seem at first sight. Glance through these and identify any with which you can agree completely:
>
> 1 A branch of the physical sciences.
> 2 A composite subject covering theories, industrial practice and analysis.
> 3 The queen of the sciences.
> 4 The sum of inorganic, organic and physical chemistry.
> 5 The work done by chemists.
> 6 A learned profession.
> 7 Not so much a subject, more a way of life.
> 8 The scientific study of matter and its transformations.
> 9 The science of substances.
> 10 The conquest of materials.
>
> ---
>
> Setting aside no. 3 (with which we hope you might agree at the end of the Course!) there is truth in all the rest. No. 4 is really a tautology, however, with the word 'chemistry' in the supposed definition. No. 7 may seem weird and we shall return to it. No. 8 sounds the most respectable and comprehensive and is often quoted to differentiate chemistry from physics (concerned with energy and its transformations); however, energetics are now so firmly a part of chemistry that the definition simply excludes too much. The remainder, in their different ways, all contribute insights into the problem, but none, on its own, is really sufficient. At the end of this Course you will be in a better position to see why.

Now it is clear that not all these answers are interpreting the question in the same way. Most of them (1, 2, 3, 4, 8 and 9) take it in its most obvious sense of *'what kind of a subject* is chemistry?' No. 1 seeks to relate it to other sciences as, by implication, does no. 8. They remind us that the nature of chemistry must be considered in terms of its place in the hierarchy of science, and, indeed, of human knowledge as a whole. On the other hand nos. 2 and 4 invite us to consider *the internal organization of the subject* while nos. 8 and 9 get as near as possible to a common theme in a few words. No. 2 is in fact taken from the title page of a well-known Victorian tome by Sheridan Muspratt, simply entitled *Chemistry* (Figure 1.4) and its reference to 'industrial practice' leads us to an alternative kind of reply.

Answers 5 and 10 are couched in operational terms, i.e. in terms of what operations are supposed to be carried out in chemistry. To assert that chemistry is what is done by chemists is remarkably unhelpful, however, since chemists are generally understood as those who practise chemistry (always excepting those

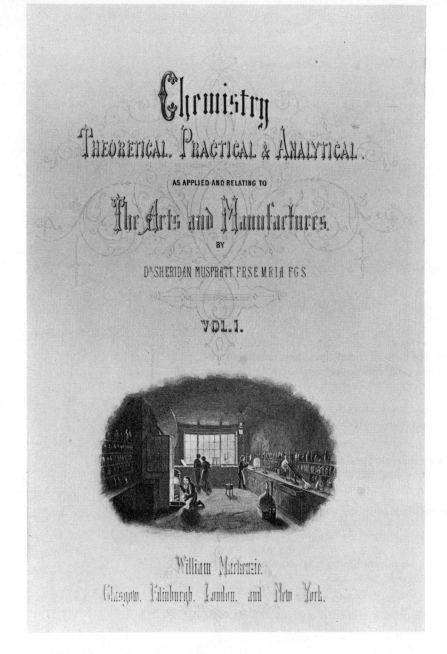

Figure 1.4 Title page from Muspratt's *Chemistry* (1857).

who work in pharmacies)! No. 10 has the merit of emphasizing the positive role of chemistry in the context of human domination; its aggressive note reminds us that chemistry is *par excellence* a useful art, and that any understanding of the nature of chemistry must take into account its *place in human history and experience*.

The answer no. 6 is, in the UK at the present time, perfectly correct, though it cannot be called a definition. Well over 100 years ago it was being argued for the first time that chemistry was becoming no longer just a subject but also a profession. This refers therefore to the *social conditions under which chemistry was practised*, another factor to be borne in mind in considering the nature of chemistry.

Finally, the light-hearted reply in no. 7 conceals two most important insights into what chemistry really is. It declares that it is not merely a subject (and therefore cannot be isolated from external factors and effects). And it also implies something about a very deeply seated attitude that usually accompanies the practice of chemistry, something which we can best call an *ideology*. It is part of our unconscious heritage from the alchemists, much more important, some have argued, than the bits and pieces of chemical technology which they bequeathed to us. As one author, M. Eliade, saw it:

> The concept of alchemical transmutation is the fabulous consummation of a faith in the possibility of changing Nature by human labours.

Some other aspects of a chemist's ideology will be apparent as the Course proceeds. All of these factors will have to be borne in mind as we examine the nature of chemistry: its ideology, the manner in which it is practised, its role in human affairs and its internal structure. In the pages that follow this last aspect will be the one that is most prominent; it is after all the most obviously 'chemical' one. But the other factors will not be neglected.

1.0.2 Objectives

At the end of this Unit you should be able to:

1 Characterize in broad terms the state of chemical knowledge in 1800.
(SAQ 1)

2 Define the *principal* stages in the development of the internal structure of chemistry from about 1800 to 1860.
(SAQs 2, 3, 15)

3 Evaluate the role of analogical argument in chemistry at that time by identifying its effects upon organic and inorganic chemistry.
(SAQs 7, 13)

4 Identify, within the context of the organic/inorganic dialogue, the chief contributions of Berzelius, Liebig, Dumas and Laurent.
(SAQs 4, 13)

5 Evaluate the applicability of the electrochemical theory of Berzelius to (a) inorganic and (b) organic chemistry in the first half of the 19th century.
(SAQs 2, 3, 8, 12, 13)

6 Evaluate the usefulness of formulations involving the radicals benzoyl, ethyl, cacodyl and 'etherin'.
(SAQs 9, 10)

7 Relate the concepts of isomerism, isomorphism and catalysis to the growing convergence between organic and inorganic chemistry.
(SAQ 15)

8 Identify the principles involved in organic elemental analysis and calculate empirical formulae from analytical data.
(SAQs 4, 5)

9 Relate the transformation of theoretical chemistry to improved techniques in elemental analysis and organic synthesis.
(SAQ 6)

10 Recognize the nature of the assumption behind:
(a) The deduction of an empirical formula from analytical data.
(b) The induction of the formula H_2O from combining volumes of gases.
(c) The rejection by Berzelius of diatomic molecules such as Cl_2.
(d) The denial (or acceptance) in the 1830s and 1840s of replacement of hydrogen by chlorine.
(e) The persistent belief in the early 19th century that organic compounds required a vital force for their creation.
(f) The ascription of similar kinds of formulae to acetic and trichloroacetic acids.
(SAQs 3, 8, 12)

11 Identify the meanings of the terms radicals, types, homologous series.
(SAQs 9, 14)

12 Criticize the thesis that unification of chemistry was significantly affected by the facts of organic synthesis.
(SAQ 10)

13 Criticize the view that there was a steady growth towards a unified chemistry from 1800 to 1860.
(SAQs 11, 15)

Table A

List of scientific terms, concepts and principles in Unit 1

Introduced in a previous Unit	Unit Section No.	Developed in this Unit	Page No.
	S100[1]		
Avogadro's 'law'	5.3.3	affinity	22
catalysis	12.4.2	alchemy	6
electrolysis	8.4.6	benzoyl radical	35
electron pair	8.4.4	cacodyl	36
free radical	13.2.1	chain reaction	42
homolysis	11.1.1	combustion analysis	26
hydrolysis	13 App. 2	copula	40
ionization theory	6.5.6	Dalton's atomic theory	14
isomerism	10.4.3	dehydrochlorination	35
mass spectrometry	6.2.5	dualism (electrochemical)	21, 33
molecular formula	6 App. 2	etherin theory	34
molecular weight	6 App. 2	homologous series	44
optical isomerism	10.4.5	isomorphism	32
oxygen theory of acids	9.10	mercury cathode	22
polarization	10.1	phlogiston	12
polarized light	10.4.5	positive halogen	42
polarity	8.4.8	pre-formation	17, 35
polymerism	13.1.2	reductionism	31
reduction	8 App. 1	regulative principle	50
sublimation	5.5	substitution, law of	39
	S24–[2]	types	44
α-substitution	4.2	vapour density	34
addition reaction	5	vitalism	21
chlorination	3.2		
distillation	9.3.2		
esterification	6.2.4		
free radical	3.2		
hydrolysis	6.2.4		
radical	3.2		
rearrangement	3.0		
solvent extraction	9.3.2		
S_N1, S_N2 reactions	3.4.2		
	S25–[3]		
allotropy	10.2		
electrochemical series	5.2		
	9 App. 1		
electron pair	9.1.4		
empirical formula	6 App. 1		
hydrolysis	6.2.9		
polarization	7.4.3		
reduction	6.1.5		

[1] The Open University (1971) S100 *Science: A Foundation Course*, The Open University Press.

[2] The Open University (1973) S24– *An Introduction to the Chemistry of Carbon Compounds*, The Open University Press.

[3] The Open University (1973) S25– *Structure, Bonding and the Periodic Law*, The Open University Press.

1.0.3 Related Course material

Home Experiment 1   See supplementary material.

TV Programme 1   This complements the Unit by discussing the way in which the organic/inorganic relationship has been affected by questions of *chemical technique* (as opposed to those of theory). It includes film made on location in Germany, as well as a number of laboratory sequences.

TV Programme 2   See also Unit 2.

Radio Programme 1   A dramatised presentation of the theme 'Controversy in chemistry' with particular reference to the atomic debates of the 1860s.

1.0.4 The nature of the problem

As we turn to the internal structure of chemistry itself we seem, at first sight, to be confronted with a simple division between organic chemistry on one hand and inorganic on the other. Physical chemistry is, of course, an extremely important area of the subject, though it stands in a different relation to the whole from the other two. The insights, techniques and results of physical chemistry are all applicable to all kinds of substances and so, in a sense, it spans them both. In the present Course physical chemistry will be constantly met with in that kind of way, rather than as a 'subject' or 'mini-subject' in itself. What will be much more obvious in the next few months will be the ways in which chemists tend to look at *substances* as belonging to either the organic or the inorganic branch, each of which has developed its own kind of approach or methodology. Indeed one of the most basic aims that we have for you is that, at the end of the Course, you will be able to appreciate something of the subtle differences that do exist today between the two branches and more precisely what their relationship is.

One thing is very obvious for a start. The organic/inorganic division is no new thing, having existed for nearly two centuries, i.e. for almost all of the 'modern' period of chemistry. Our purpose in Units 1–3 is to explore in some depth how the present relationship has come about.

Today this theme is especially important because the role of organic and inorganic chemistry in the total structure of the subject is a matter of urgent and sometimes passionate debate. The underlying causes of the current tension will be examined in Unit 3, although they will be partially exposed in Units 1 and 2 as well. A fairly representative plea for a new approach was made in 1969 by Professor G. S. Hammond:

> For several years, I have been saying that it is a mistake to describe chemistry using a conceptual organization that was created in the last century. The meanings of the words that we use—organic, inorganic, physical and 'analytical'—are for the most part obscure to anyone who has not already been thoroughly initiated in the ritual of chemistry. Organic chemistry certainly sounds like the chemistry of organs and inorganic chemistry should deal with nonorgans. Since chemistry is recognized as a physical science, the term 'physical chemistry' seems entirely redundant. Analytical chemistry is a little easier to identify, but one is confused to learn that only a minority of people who call themselves analytical chemists regard chemical analysis as their principal research interest. Furthermore, even the traditional meanings of the names known to the ingroup of educated chemists are not really good descriptions of coherent fields of reactivity. For example, good work in photochemistry is done not only by organic, inorganic, physical and analytical chemists, but also by biochemists and biophysicists. My colleague, Professor Harry Gray, is usually called an inorganic chemist, but he states that, 'Inorganic chemistry is a ridiculous field.'

By the time you have finished these three Units you should be in a good position both to assess the historical accuracy of some of the statements in that quotation and to form your own view as to the novelty of its approach. Naturally the Course Team has faced this issue (among many others) while designing the Course and has come to certain conclusions. What these are will become clearer as the Course proceeds, and after the final Unit you should be in a position to say whether or not you agree with our decisions, and to have your own opinion on the remarks of Professor Hammond above.

To alert you to the kind of problems involved we are giving below another quotation from the same conference, coming as part of the discussion on Hammond's paper. These are some remarks made by Professor G. Illuminati of Rome. Read them carefully and then attempt the ITQ immediately following.

> To the extent that there exists such a thing as chemistry itself, all internal branches of chemistry have developed because of an artificial classification of Nature due to a first approximation process in our knowledge of science. Objectively, there are no such things as inorganic, organic or physical chemistry except in our minds. By classifying science in this way we introduce a distortion of Nature. The more we know about chemistry the more we do not need to divide it in what has been called the traditional way. So not only ought we to make every effort to update our teaching by presenting chemistry as we see it *today*, but also we should make it clear to ourselves and to the students that the most honest presentation of science is fundamentally motivated by the present objectives of studies as indicated by the trends of interests which are bound to change, as they should, from time to time.

(a) Illuminati appears to object to the traditional presentation of chemistry as inorganic, organic and physical on two rather different grounds (though they are not clearly separated in his argument). What are they?

(b) What assumptions are concealed in these two objections?

(a) The objections to the traditional approach are:

(i) That it is *artificial*: 'there are no such things as inorganic, organic or physical chemistry except in our minds.' It is thus a distortion of nature.

(ii) That it is *antiquated:* chemistry must be presented 'as we see it *today*.'

(b) The assumptions in these two objections seem to be as follows:

(i) The claim that it is artificial to present chemistry in its traditional three divisions assumes that there is a coherent thing ('chemistry') to divide in the first place. In other words, it assumes that dividing science into physics, chemistry, biology, etc., is more praiseworthy than dividing chemistry further. But it could be argued that the very existence of physical chemistry has dissolved the boundary between physics and chemistry and that neither of these has a right to an independent existence; similarly for biochemistry, etc. The speaker virtually concedes this point in his opening words. His objection also implies that other divisions would be *less* artificial. It is hard to see why this should be so for any other reason than that they are more contemporary. This brings us to the second objection.

(ii) The assumption behind the second objection is clearly that what is latest is best. It has often been argued that without this belief science could not progress, scientists would be out of a job, and so on. In the sense that we now have the most powerful techniques, the accumulated data of past and present chemistry and (one hopes) the accumulated wisdom to deal with it, one cannot fault this objection. Was it not Newton who said that he saw further than the giants who preceded him because he stood upon their shoulders? Yet we must not confuse scientific facts and theories with their presentation. It does not *necessarily* follow that new theories cannot be fitted into an existing framework, perhaps with some modification. To argue that chemistry must be viewed in fundamentally new ways because of our 'trends of interests' looks suspiciously like replacing one artifice for another.

You might be amused to compare these sentiments with the following quotation from a paper by E. W. Mills in 1871. It will at least illustrate that it is no new thing to attempt a reformation of the structure of chemistry on the grounds that the present structure is artificial ('untrue to nature') or just old. Mills was one of a group of enthusiasts anxious to reform chemistry by getting rid of atoms. You will appreciate that modern chemists would find great difficulty in recognizing a subject that had been restructured along the lines he proposed.

> The atomic theory...readily blends with all the prejudices of our education and is reinforced by them; so that after some years it becomes an essential part of the mind, which has no longer the power to reject it, even with the aid of the desire...A logical mind...will find, if my argument be sound, that the atomic theory has no experimental basis, is untrue to nature generally, and consists in the main of a materialistic fallacy, derived from appetite more than from judgement.

In studying the changing nature of the interface between organic and inorganic chemistry we shall not be delving into the past for the sake of it but to seek for a genuine explanation for the shape of chemistry today, and possibly to gain a better appreciation of current developments. This will mean taking advantage of the substantial body of recent scholarship by a number of chemical historians. For some reason chemistry has been a more popular theme for historians of science than much else, but it has to be admitted that a good deal of this writing has limited relevance for us for two reasons: on one hand it tends to finish too early, often within the 19th century; on the other it may well be not sufficiently *chemical*, overlooking obvious issues of chemical importance in the search for erudite alternative theories for the course of historical events. However, there are notable—and noble—exceptions to this generalization in recent articles, papers and books.

Our purpose now is not to present a rehash of such work, though the debt to this will be considerable. It is rather to offer a new model for the relationship between organic and inorganic chemistry over the whole of the period, emphasizing at the same time that this is simply a framework into which to fit the facts, just like a strictly chemical theory and just as susceptible to test and rejection.

The matter can be put rather simply like this. Throughout the period under review there have been two opposite tendencies at work, the one being to unite the two branches of chemistry, the other to force them apart. Each of these has been a resultant of a complex of factors of which the magnitude has varied with time. Consequently, at different periods, the two branches have seemed to be converging towards each other, at others to be diverging apart. We have in fact a series of phases, alternately convergent and divergent, marking the progress of chemistry up to our own time. It seems possible to identify six of these and in examining each of them we shall learn a great deal about the factors that have gone into the evolution of chemistry to its present form. These stages were not marked by hard-and-fast boundaries; the end of one tended to overlap the start of the next. They may be briefly indicated as follows, together with their place in these Units:

I	Recognition: up to the early 19th century	Divergent	
II	Incorporation: up to about the 1840s	Convergent	Unit 1
III	Secession: 1840s and 1850s	Divergent	
IV	Integration: predominantly the 1860s and 1870s	Convergent	Unit 2
V	Specialization: 1880s to 1940s	Divergent	
VI	Unification: perhaps the tendency of today?	Convergent	Unit 3

1.1 The recognition of organic chemistry—PHASE I

1.1.1 The quest for new substances

To understand the shape of present chemistry we have to go back in time nearly 200 years. We have said this several times already, and now we are in a position actually to do so. Oddly enough the more we know of modern chemistry the harder this is likely to be at first. Those who are really immersed in research at the frontiers of chemistry may find the task almost impossibly difficult since a prerequisite is the ability to think in a sympathetic way about attitudes that we find unacceptable today, even 'wrong' by modern standards. Nevertheless we must try. In evaluating early work we must take no account of such fundamental concepts as structure, or even more basic ones still such as valency or atoms. You will note that we must *take no account* of them; it is stupid to assert (as some have done) that we must *forget* them, for no one worthy of advanced chemical study could ever liberate his memory from such totally essential categories of thought. What we need now is imagination, not pretence. In that spirit we turn to the state of chemistry shortly before 1800.

We should just be in time to witness the last funeral rites for that terror of generations of schoolboy chemists, the theory of phlogiston.* In a few places its spirit

*i.e. the belief that combustion involved the giving to the atmosphere of a substance, phlogiston, present in all combustibles.

lingered on into the new century but effectively it was dead. There is no need to exhume it now, though in its lifetime it had been a great integrating principle of chemical philosophy. Through the work of Priestley, Scheele and (above all) Lavoisier one could now explain combustion and respiration in terms not of a phlogiston given off into the atmosphere, but rather as a combination with a new element—oxygen—present in the air. The other main constituent of this was nitrogen, or azote as the French preferred to call it. And oxygen was also present in a combined form in water, which now stood fully revealed as a compound of hydrogen. The whole question of what an element really was became settled with a new finality in Lavoisier's definition 'the last point which analysis was capable of reaching'. By this he meant that any substance incapable of being chemically broken down into simpler substances must be regarded as an element (Figure 1.5). His list of such elements (given here in modern symbols) included many familiar to us today:

Non-metals: O, N, H, S, P, C, Sb, As.

Metals: Ag, Bi, Co, Cu, Sn, Fe, Mn, Hg, Mo, Ni, Au, Pt, Pb, W, Zn.

It also included certain materials that we know to be oxides but which analysis had *not* then been able to reach (notably oxides of calcium, magnesium, barium, aluminium, silicon). Heat and light (*calorique* and *lumière*) were included, as were certain 'radicals'—suspected fragments of substances but not as yet isolated. Muriatic and fluoric radicals (in our chlorides and fluorides) corresponded to the halogens chlorine and fluorine, still unknown, while the boracic radical was apparently an oxide of boron.

	Noms nouveaux.	Noms anciens correspondans.
Substances simples qui appartiennent aux trois règnes & qu'on peut regarder comme les élémens des corps.	Lumière............	Lumière.
	Calorique.........	Chaleur.
		Principe de la chaleur.
		Fluide igné.
		Feu.
		Matière du feu & de la chaleur.
	Oxygène.........	Air déphlogistiqué.
		Air empiréal.
		Air vital.
		Base de l'air vital.
	Azote............	Gaz phlogistiqué.
		Mofete.
		Base de la mofete.
	Hydrogène.......	Gaz inflammable.
		Base du gaz inflammable.
Substances simples non métalliques oxidables & acidifiables.	Soufre...........	Soufre.
	Phosphore........	Phosphore.
	Carbone,.........	Charbon pur.
	Radical muriatique.	Inconnu.
	Radical fluorique	Inconnu.
	Radical boracique,.	Inconnu.
Substances simples métalliques oxidables & acidifiables.	Antimoine........	Antimoine.
	Argent...........	Argent.
	Arsenic..........	Arsenic.
	Bismuth..........	Bismuth.
	Cobolt...........	Cobolt.
	Cuivre...........	Cuivre.
	Etain............	Etain.
	Fer..............	Fer.
	Manganèse.......	Manganèse.
	Mercure..........	Mercure.
	Molybdène.......	Molybdène.
	Nickel...........	Nickel.
	Or...............	Or.
	Platine..........	Platine.
	Plomb...........	Plomb.
	Tungstène.......	Tungstene.
	Zinc.............	Zinc.
Substances simples salifiables terreuses.	Chaux...........	Terre calcaire, chaux.
	Magnésie.........	Magnésie, base du sel d'Epsom.
	Baryte...........	Barote, terre pesante.
	Alumine.........	Argile, terre de l'alun, base de l'alun.
	Silice...........	Terre siliceuse, terre vitrifiable.

Figure 1.5 Page from Lavoisier's *Traité* (1789).

In all of this the now familiar chemical atom was conspicuously absent, as of course were molecules, valencies and structure. It might understandably be asked, in that case, what on earth did chemists do with their time in the next few years? The shortest (and probably best) answer is 'discover new substances', though all kinds of other research were of course in hand. But this was a great period of expansion and exploration. In what territories this took place we shall now discover.

The late 18th century was, understandably, a great time for discoveries in the chemistry of gases—*pneumatic chemistry* as it was called. The experiments of the aristocratic English chemist Henry Cavendish on 'inflammable air' established hydrogen as a definite chemical individual, while in Scotland Joseph Black was clearing up the relationships between carbon dioxide ('fixed air') and the carbonates, being the first to recognize a gas as a different chemical substance from air. The Swedish chemist Scheele discovered 'more new substances of fundamental importance than any other chemist without exception' (J. R. Partington), including oxygen (independently of Lavoisier), chlorine, arsine, hydrogen fluoride, silicon tetrafluoride, hydrogen cyanide and many other substances.

The early 19th century saw inorganic chemistry make remarkable strides forward in the isolation of *new elements*. Table 1 indicates the dates of most discoveries for the first 60 years of the century.

Figure 1.6 C. W. Scheele (1742–86), the Swedish chemist who discovered oxygen before and independently of Priestley.

Does any pattern emerge from this Table?

Right at the beginning of the 19th century there was a cluster of discoveries associated with the platinum group of metals then arousing much interest, partly because of increasing availability of platinum ores. Then the discoveries of 1807–8 arose from the recently discovered phenomenon of electrolysis and its exploitation by Humphry Davy. There was then a relative lull for ten or so years. Much of the progress until 1830 was in Sweden, reflecting a long Swedish tradition in mineralogy and the richness of the ores in that land. In much of this work we can discern the achievements and influence of one of the greatest chemists of the century, Jöns Jacob Berzelius. Elements isolated by him or his co-workers were Ce, Li, Se, Zr, Ti, Th, V. We shall meet Berzelius frequently in these Units. Finally there is a perceptible falling-off in the rate of discovery of the elements— actually until the 1860s.

The value of lists like this is strictly limited, but one can at least infer that after an initial spurt the rate of discovery became remarkably slow. One of the reasons, as we shall see, was that the chemical world was becoming preoccupied with quite different matters. In what we now term inorganic chemistry there tended to be a strong concentration on analysis (see TV programme 1). Also the atomic theory of John Dalton, published by stages in the first ten years of the century, was focusing attention on combining quantities (or 'equivalents') and raising tantalizing questions about whether all atomic weights should be whole numbers and, if so, whether all elements might not be made of hydrogen. There was until well into the second half of the century no Periodic Table and no concept of valency; in consequence, this kind of chemistry lacked coherence and interest. It also tended to lack relevance; admittedly the industrial processes associated with the manufacture of acids and alkalis were being rapidly developed, but often their links with chemical theory were pretty tenuous. Much the same could be said of the infant science of metallurgy.

Given this rather dismal picture of chemistry in the early 19th century we return to our assertion that this was to be a great period of expansion and exploration; it was not so in the areas of traditional interest to chemists, so where was it to be?

1.1.2 Early organic chemistry

A glance at any textbook of this time (i.e. shortly after the French Revolution) will immediately reveal the areas being explored. Such books were almost always organized according to the sources of the materials described, rather on the lines of a modern 'Twenty Questions': animal, vegetable and mineral. Of these three the last many times exceeded in extent the other two, which tended to be lumped together as being rather different in originating in living organisms. In 1675 N. Lémery in his *Cours de Chymie* pointed out the distinction between mineral and vegetable bodies, and in 1777 the Swedish chemist Bergmann was heard to

TABLE 1

Year					
1800					
1801					
1802	Pd				
1803	Ce				
1804	Os	Rh	Ir		
1805					
1806					
1807	K	Na			
1808	Ba	Sr	Ca	Mg	B
1809					
1810					
1811	I				
1812					
1813					
1814					
1815					
1816					
1817	Li	Cd			
1818	Se				
1819					
1820					
1821					
1822					
1823					
1824	Zr				
1825	Ti				
1826	Br				
1827	Al				
1828	Be				
1829	Th				
1830	V				
1831					
1832					
1833					
1834					
1835					
1836					
1837					
1838					
1839	La				
1840					
1841	U				
1842					
1843					
1844	Ru				
1845					
1846					
1847					
1848					
1849					
1850					
1851					
1852					
1853					
1854					
1855					
1856					
1857					
1858					
1859					
1860	Cs				

speak of 'non-organic and organic bodies' and so to give expression to a distinction that has survived in some form to this day. Certainly it was around this time that it was first clearly recognized.

Let us be clear as to what this distinction was really about. It was not about structure, properties or abundance—at least not at first. It was simply a matter of origin. No one doubted then that there was a fundamental difference between living and non-living things (though not all would have agreed as to exactly where the boundary should be drawn). So it was natural to classify substances according to source, though this was not always as easy as it might seem. For example, was honey *vegetable* (from flowers) or *animal* (from bees)? By the beginning of the 19th century there was a widespread, if vague, recognition of organic substances as in some ways different from the rest. This marks clearly the first phase of fragmentation within chemistry.

To understand what happened next it is rather important to realize the scope of the problem. The surprising thing to a modern chemist is the small number of organic substances then recognized. Table 2 lists a couple of dozen of these substances by modern formulae. It is probably not quite complete but there are very few important omissions. Several complex substances which we know today as mixtures are not included (such as blood, fats, plant oils, gum, etc.).

By way of revision, fill in the missing names in Table 2.

4	acetic acid	12	acetone
5	ethanol	15	acetaldehyde
6	diethyl ether	16	glycerol
7	ethyl nitrate	17	oxalic acid
8	chloroethane	18	malic acid
9	benzoic acid	19	tartaric acid
10	succinic acid	22	urea
11	formic acid	24	1,2-dichloroethane

This must seem an odd assortment of chemicals. To see why these should be the ones discovered, and not others, will be the object of the next question and of a Home Experiment.

Of the six substances listed for the 16th century, four are closely related in structure. (a) Which four are they? (b) How could they have been obtained? (c) What experimental technique would have been a prerequisite for their production and isolation?

(a) Nos. 5, 6, 7 and 8 have the obviously common feature of an ethyl group.

(b) Nos. 6, 7 and 8 could be prepared from ethanol by the actions of (respectively) fairly concentrated sulphuric, nitric and hydrochloric acids. However, the constitutions of the products were totally unknown; it was believed until well into the 19th century that 'ether' contained sulphur (the impure samples often available probably did!).

(c) To obtain these acids and fairly pure ethanol the technique of *distillation* would have to have reached a fairly high level of efficiency. This is exactly what did happen about this time, largely as a result of the activities of the alchemists. It may be observed, however, that completely anhydrous ethanol did not become available until just before 1800. Several other products in the list (notably nos. 15 and 24) were later derived from ethanol.

To convey some feel of the situation a simple experiment has been devised. If possible attempt this now. It is HE 1 in the *Home Experiment Book*.

In the light of that experience you should now look at Table 2 again.

How many of these substances would be expected to be water-soluble?

Most of these will dissolve in hot or cold water, the chief exceptions being the two dyestuffs (2 and 3), and the artefacts from ethanol (6, 7, 8 and 24). The occurrence of a high proportion of OH groups in most of the molecules should have been a sufficient clue. A few others as well, e.g. acetone and acetaldehyde, will form hydrogen bonds with the OH groups in water.

TABLE 2

1	$C_{12}H_{22}O_{11}$	sucrose
2	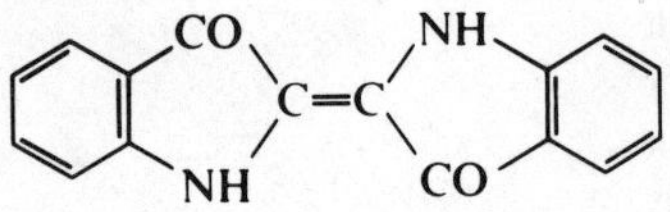	indigo
3	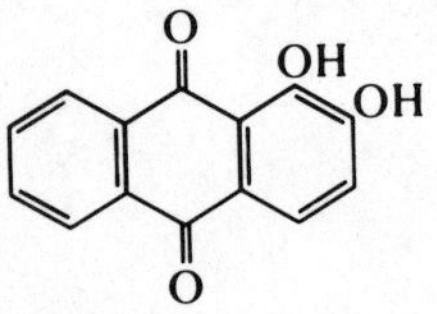	alizarin
4	CH_3COOH	

Ancient World (brackets 1–4)

5	C_2H_5OH	
6	$(C_2H_5)_2O$	
7	$C_2H_5ONO_2$	
8	C_2H_5Cl	
9	C_6H_5COOH	
10	$\begin{array}{l}CH_2COOH\\ |\\ CH_2COOH\end{array}$	

16th Century (brackets 5–10)

11	$HCOOH$	
12	CH_3COCH_3	
13	$C_{12}H_{22}O_{11}$	lactose
14	$C_6H_{12}O_6$	glucose

17th Century (brackets 11–14)

15	CH_3CHO	
16	$\begin{array}{l}CH_2OH\\ |\\ CHOH\\ |\\ CH_2OH\end{array}$	
17	$(COOH)_2$	
18	$\begin{array}{l}CH_2COOH\\ |\\ CH(OH)COOH\end{array}$	
19	$\begin{array}{l}CH(OH)COOH\\ |\\ CH(OH)COOH\end{array}$	
20	$\begin{array}{l}HOOCCH_2C(OH)COOH\\ \quad\quad\quad |\\ \quad\quad CH(OH)COOH\end{array}$	citric acid
21	$C_6H_{10}O_8$	mucic acid
22	$CO(NH_2)_2$	
23	$C_5H_4N_4O_4$	uric acid
24	$ClCH_2CH_2Cl$	
25	$C_{27}H_{45}OH$	cholesterol

18th Century (brackets 15–25)

SAQ 1 (*Objective 1*) Given the materials available in 1800, which of the following experiments *could have been* carried out in that year?

(a) Nitration of benzene

(b) Bromination of ethylene

(c) Formation of ethyl benzoate

(d) A Grignard reaction

(e) Preparation of oxalic acid by oxidation

1.1.3 A new expansion

Thus, in summary, it would appear no accident that the organic compounds known in a fairly pure state at this time were those that could be readily isolated by water-extraction or by sublimation. In his *Système des Connaissances Chimiques* (*System of Chemical Knowledge*) of 1800 the French chemist Fourcroy maintained that a crucial factor in the study of organic compounds was a realization of their sensitivity to heat. As you have found, they readily decompose in a complex manner. Until the mid-18th century many chemists believed that the products of heating were already present in the original material ('pre-formation'). In other words, heat did not effect a chemical decomposition as we understand it but only a separation. When the same product turned up from very different starting materials this belief was weakened (compare the 'caramel' production from both sucrose and tartaric acid). In any case, less drastic methods of extraction were called for, and the use of solvents, particularly water, became common. Much pioneer work of this kind was done in France during the late 18th century. With the additional use of organic solvents the number of new materials extractable without decomposition began to increase rapidly and so did the number of pages needed to describe them in the textbooks. A good example of the scale of the advance in the early 19th century is the multi-volume *Larbök i Kemien* or *Textbook of Chemistry*, by J. J. Berzelius. The details of the first Swedish edition are as follows (it was also published in French and German):

Vol. I,   1808: inorganic chemistry
Vol. II, 1812: inorganic chemistry
Vol. III, 1818: vegetable chemistry and chemical proportions
Vol IV,   1827: vegetable chemistry
Vol V,   1828: vegetable chemistry
Vol VI,   1830: animal chemistry

It is not surprising that by 1829 he could write of 'the new science of organic chemistry'.

But what could justify such a separation from the salts, oxides and minerals studied in the inorganic realm? One factor was the sheer number of the organic bodies, from a few dozen to many hundreds in two or three decades. Moreover, they shared many characteristic properties—solubility in solvents like alcohol, low melting temperature, thermal instability, etc. Their constitution was at first an open question but they did seem, on the whole, much more complicated than the inorganic substances. But above all there was the question of *origin*. Today we are so used to discounting this as a factor of importance that it really is hard to take it seriously. For us sugar is sugar, for example, no matter where it comes from; its structure is the important thing. But no one in 1820 was to know that. Hence the mysterious nature of life itself was quite sufficient justification for regarding its chemical products as a group on their own. The other differences merely reinforced this impression.

1.2 Incorporation of organic chemistry—PHASE II

1.1.2 The state of the problem

By about 1810 organic chemistry had emerged as a recognizable subject in its own right. The name has stuck ever since. But the pressing question that faced chemists then, just as in a different form it still faces us today, was the relationship between the new science and the old. In practical terms this meant the relation between organic and inorganic chemistries.

To avoid going into immense historical detail, but at the same time to give some idea of the 'feel' of the situation, we are going to examine one statement in some depth. This comes from a paper by the Swedish chemist J. J. Berzelius. Brief biographical details will be found in the caption to his portrait (Figure 1.7).

Figure 1.7 Jöns Jacob Berzelius (1779–1848), one of the 19th century's greatest chemists. Born and working nearly all his life in Sweden he entered chemistry from medicine, becoming preoccupied with electrolysis at about the same time as Davy (from around 1802). Hearing shortly afterwards of Dalton's atomic theory he fused the two concepts of atomism and electrochemical dualism into a system of chemistry that, for all its failings, gave a unity and direction to chemistry that it has never wholly lost. He also discovered cerium (1803), selenium (1817) and thorium (1829), clarified the concepts of isomerism and catalysis, devised the term 'protein', personally determined atomic weights of all but a handful of the elements then known, and invented the modern alphabetical system of chemical notation. His theoretical ideas were powerfully advocated in his *Jahresberichte* (annual reports), his multi-volume textbook in six languages (though not in English) and by other books, including his 1819 *Essai sur la Théorie des Proportions Chimiques*, and, of course, a large output of research papers. Few men since Lavoisier have so profoundly affected the course of chemical history as did Berzelius.

Before we do look at the quotation we should perhaps make the obvious point that to concentrate on one man's writing can be dangerous to the extent that that man may not be truly representative of his time. However, if one is aware of the problem one can make corrections for it, and this we shall attempt to do. In any case Berzelius, perhaps more than any other man, did determine how many other chemists thought; he set trends instead of merely reflecting them.

The passage below comes from a paper in the French periodical *Journal de Physique* (1811, **73**, 260–1). The issue in which the paper appears contains also the first announcement of the epoch-making hypothesis of Avogadro, which, together with Berzelius' offering, must make it one of the most important single issues of a journal in the whole of chemistry. Berzelius himself was concerned to rationalize the chaotic chemical nomenclature of the day. This he did on the basis of systematic Latin names, so giving the world an international language, comparable in its universality to musical notation and the use of mediaeval Latin. It may be noted in passing that, in 1814, he also introduced a modern chemical symbolism, with letters standing for atoms of the elements.

Names were given to substances on the basis of their chemical nature, and they were classified according to their behaviour in electrolysis. This led Berzelius to produce the first electrochemical series of the elements. But this is all incidental, for the passage that is our immediate concern contains some startling ideas about the nature of organic compounds. Compared with modern chemical writing it may seem to lack precision and conciseness, but in those respects it reflects the state of science at the time. In fact, however, Berzelius was constantly injecting into chemistry ideas of much greater precision than his contemporaries would wish. The passage reads as follows:

> These [organic bodies] are composed of two or sometimes several combustible bodies combined with a quantity of oxygen, which is common to them, and is usually sufficient only to oxidize one of these combustibles. Their constituent parts cannot be separated without combining together in several new ways to produce binary combinations and to share the oxygen between them according to their very complicated affinities. Elements which constitute organic bodies obey the same general law as that regulating the formation of inorganic combinations; but the innumerable varieties which this law admits in organic nature are not yet sufficiently clearly defined...
>
> ...The chief condition for organic formation seems to be an electrochemical modification in the elements which differs from that which they normally have in inorganic nature. After they are removed from the organic body which has produced them they tend always to regain this original electrochemical modification; it is a result of this tendency that they ferment and decompose on contact with air and water and on exposure to a high temperature.

In the light of this passage please attempt short answers to the following questions. You can then compare these with the suggestions given below, all of which will be developed in the discussion that follows:

1 How were the compositions of organic bodies imagined? Is this acceptable today?

2 What common feature linked organic and inorganic chemistry?

3 What *specific* problems were presented by the formation of organic compounds?

4 What was a prerequisite for 'organic formation'?

---

1 As compounds of *combustible bodies* (C and H) with oxygen. They were *tertiary* or *quaternary* formations, i.e. were made of 3 or 4 elements.* Clearly a modern chemist would be unhappy about a scheme which excluded hydrocarbons (Berzelius tended to regard these as inorganic—very few were then known) and amines, to say nothing of halogen and sulphur compounds, etc.

* The further distinction was often made that compounds of vegetable origin were tertiary (C, H and O) while those from animals were quaternary (C, H, O and N). Once analytical techniques were established such a view could not survive. It disappears early in the 19th century.

2 Both groups obeyed the *same general law*. This is not spelt out further, but Berzelius, as we shall see, has dropped a pretty broad hint that chemistry ought to incorporate the new science rather than simply coexist with it.

3 The fact that chemistry, by itself, can rarely produce organic bodies from inorganic. Hence a living body is needed.

4 Something he designated rather oddly as *an electrochemical modification in the elements*. More of this later.

We shall now develop these themes in rather more detail.

1.2.1.1 The composition of organic compounds

In regarding organic compounds as always containing oxygen Berzelius was drawing on the idea proposed 30 or 40 years earlier by Lavoisier that vegetable substances contained carbon, hydrogen and oxygen while animal substances contained these elements together with nitrogen and possibly sulphur, phosphorus, etc. Thus Lavoisier, who had discovered oxygen and for whom oxygen was the all-important element, regarded organic substances as oxides of compound radicals. Oxygen, for him, was the principle of acidity (all acids then known being derived from non-metallic oxides). So, given enough of it, one could convert a neutral oxide into an acidic one.

See S100, Unit 9

The discovery, in 1800, that the neutral carbon monoxide could be oxidized to the acidic dioxide lent support to this view. An organic example already mentioned is

$$C_{12}H_{22}O_{11} + 18[O] \longrightarrow 6(COOH)_2 + 5H_2O \quad \text{(modern symbols)}$$

sugar   oxalic acid

Further oxidation converts the oxalic acid to CO_2.

At this stage, of course, such symbols could not be applied since no one knew that formulae could be written, let alone what they were. But in an age aglow with a new understanding of what combustion was all about it was entirely reasonable to think of organic compounds as much more complex oxides. And this brings us to our next point.

1.2.1.2 The birth of an analogy

Lavoisier had, perhaps unconsciously, permitted himself to draw an analogy between (say) the oxides of sulphur and the organic acids (note how many of the items in Table 2 fall into that category). Now Berzelius spells this out more explicitly. To assert the reign of one general law over both areas is to establish some relationship between them—however vague the law may be. As we shall see before the end of this Unit, it was this analogy—between organic and inorganic compounds—pursued and extended to an amazing degree, that led to a strong convergence between the two branches of chemistry and to the virtual incorporation of the new one within the framework of the old.

A further point that arises from this is that it was for Berzelius an act of faith and intuition that led to the propounding of the analogy. He realized the fragmentary nature of his knowledge but asserted that nevertheless the same general law applied. This is in no sense a denigration of either Berzelius or of chemistry. The simple fact is that science often progresses on the strength of such hunches as this. We shall see several more examples in the weeks ahead. In some cases what begins as an act of faith turns out to look suspiciously like the actual truth as new facts emerge; in other cases it just becomes an unquestioned presupposition of scientific dogma. It may be hard for those immersed in research to see this at the time, but it is no bad thing for all of us to reflect occasionally on the logical status of all our beliefs. We hope you will do this yourself as the Course progresses.

20

1.2.1.3   The uniqueness of the life process

Berzelius, like many others before and after him, was puzzled by the inability of the chemist to emulate a living organism and produce in the laboratory such complex substances as, for example, sucrose (actually this was not synthesized until 1953, by Lemieux and Huber). The general belief that only life can produce organic compounds is sometimes loosely labelled *vitalism*; this is an area strewn with misconceptions, and it is only for that reason that we must briefly glance at it here. One point sometimes made is that Berzelius and his kind were here taking refuge in the occult, shrinking from a view of science in which all would be reduced to the simple categories of physics and chemistry. Rather than immediately banishing Berzelius to the backwoods, however, it would be as well to heed what he actually said; he did *not* assert an unequivocal vitalism and he did offer a very tentative explanation in terms of electrochemistry. Moreover, neither he nor most other chemists for decades to come regarded the issue in terms of *reductionism*—reducing life to physical levels of explanation. They were talking about organic compounds, not organized bodies; there was—and is—no logical reason why the elimination of vitalism from chemistry should automatically eject it from physiology. There may be other reasons but they have nothing to do with organic chemistry.

A further misconception that you are almost certain to meet is this: vitalism provided the rationale for the separation of organic from inorganic chemistry; hence its eventual demise would mean a union of the two branches.

As we shall see (Unit 2) the slow extinction of chemical vitalism had very little to do with the changing organic/inorganic interface; other far more important forces were at work. But you may have read that vitalism was extinguished by Wöhler's preparation of urea in 1828. It is true that he wrote to Berzelius 'I can make urea without the necessity of a kidney, or even of an animal, whether man or dog', but he went on immediately to add 'ammonium cyanate is urea'. For him, as for others, the great point of interest was in what we should call *isomerism*. But the claim that the elimination of chemical vitalism was begun—or even accomplished—by Wöhler's synthesis of urea in 1828 is astonishing only for its brazen repetition despite a multitude of accumulated evidence to the contrary. To mention but one aspect, urea was, by the criteria then in vogue, a highly doubtful contender for a place in the organic hierarchy. It was (rightly) regarded as a decomposition product of organic material, the end-product of nitrogen-metabolism, and Wöhler's starting-point (NH_4CNO) was ultimately obtained from hooves and horns, organic enough for most people! Perusal of the chemical literature of the time will demonstrate that chemists were unaware of the 'drama' of Wöhler's discovery that posterity has invented for it. Over 30 years ago one scholar remarked 'those who believe that Wöhler drove vitalism out of organic chemistry will believe anything' (D. McKie (1944) Wöhler's synthetic urea and the rejection of vitalism, *Nature*, **153**, 608). It is of, course, a fact that 'bodies composed according to the principle of organic formations' (as Berzelius called them) *did* become available by chemical synthesis as the century wore on, but this played no part whatever in the coming together of the two branches with which we are now concerned.

1.2.1.4   An electrochemical view of matter

Berzelius asserted that in organic compounds the elements of which they are composed have undergone an 'electrochemical modification'. Neither here nor elsewhere did he ever indicate specifically what this was; he simply did not know. The phrase is important, therefore, not for what it tells us about the special characteristics of organic compounds but rather for what it indicates about Berzelius' overall approach to chemistry. This has an importance for the future nature of the subject that it would be hard to exaggerate. What was this special approach?

In a word, Berzelius looked upon all chemistry in the light of the fact of electrolysis. This phenomenon had been discovered (quite by accident) by Nicholson and Carlisle in 1800 and within the next ten years had been vigorously exploited by Humphry Davy in England. Not the least remarkable outcome had been the isolation of the alkali and alkaline earth metals.* At one stage, in fact, Davy was

Figure 1.8 Friedrich Wöhler (1800–82), German chemist, once a pupil of Berzelius and, from 1836, Professor at Göttingen. He isolated metallic aluminium and beryllium, discovered the isomerization of ammonium cyanate and, with Liebig, investigated a series of reactions involving the benzoyl radical.

---

* Davy's experiment is described in S25–, Unit 3, and reproduced in A202, TV programme 10 (together with that of Nicholson and Carlisle).

Figure 1.9 Humphry Davy (1778–1829), English chemist, famous above all for his discovery of sodium and potassium (1807, by electrolysis) and his invention of the miners' safety-lamp (1815). He also pioneered nitrous oxide anaesthesia, while his experiments on electrolytic phenomena justly entitle him to be regarded as the founder of electrochemistry; his electrical conception of chemical change was taken up by Berzelius. It has been said that Davy's greatest discovery was his young assistant, Michael Faraday.

indebted to Berzelius, whose discovery of the use of a mercury cathode (1808) enabled his English colleague to make further advances. But in theoretical terms the debt was the other way round. At an early stage Davy had been led by the remarkable facts of electrolytic decomposition to seek further relations between chemical and electrical changes. In his conclusion, 'chemical affinity' means roughly the holding together in chemical combinations:

> Is not what has always been called chemical affinity merely the union or coalescence of particles in naturally opposite electrical states? And are not chemical attractions of masses owing to one property and governed by one simple law?

These words, spoken in 1808, mark the effective beginning of a view that has persisted to this day, that chemical changes are essentially electrical in nature. From this springboard Berzelius was able to launch himself into the murkiest depths of chemical theory. The rest of his days were to be spent in efforts to bring the whole of chemistry into the light of his electrochemical theory. As he said in the 1811 paper from which we quoted in Section 1.2.1:

> In electricity there is an agency so powerful that we are beginning to suspect that this agency is identical with chemical affinity. The different relations of bodies with respect to electricity will henceforth be the basis of all chemical systems.

Later he was to modify this simple conception in many sophisticated ways. But to the end he maintained that chemical affinity was basically electrical in nature and that elements in compound bodies were held together by electrical attraction.

At the foundation of Berzelius' theory lay his arrangement of all the elements into one comprehensive series 'relative to their electrochemical properties'. At one end came oxygen 'the most electronegative of all bodies' and at the other the highly electropositive metal potassium. In between came all the rest, ranked according to the extent of their electropositive or electronegative character. This was a reflection of their behaviour during electrolysis, electronegative substances going to the anode and electropositive to the cathode. Sometimes a metal would not be deposited but its oxide might instead collect around the cathode; in such cases the assumption was that 'the electrical order of combustible bodies in general agrees with that of their oxides'. He does not give details as to how all the placings were arrived at, but speaks of several factors which could help. One was the phenomenon of contact electrification which had so impressed Humphry Davy before him; this was the development of charges of what we should call static electricity when two dry substances were placed in contact. Thus solid acids on plates of copper become negative, but bases cause the *metal* to be negative. Lime and oxalic acid on contact become respectively positive and negative. This was an extension of Volta's discovery that dissimilar metals become electrified when they touch each other—another fact which Berzelius seems to have used. Again the acidic or basic character of an element's oxide was a guide to its degree of electronegative character. The one obvious yardstick that he seems to have rejected is that of a displacement series for metals; thus we know that a metal may be replaced in its salts by one more electropositive, as:

$$Zn + CuSO_4 = Cu + ZnSO_4$$

Possibly, as in ease of combustion in oxygen, he thought there might be too many complicating factors for this to be a reliable method. However, he proceeded to obtain what we might call an electrochemical series. Table 3 gives one version of this (from 1831); the symbols are modern:

TABLE 3

O S N F Cl Br I Se P As Cr Mo W B C Sb Te Ta
Ti Si H Au Os Ir Pt Rh Pd Hg Ag Cu U Bi Sn Pb
Cd Co Ni Fe Zn Mn Ce Th Zr Al Y Be Mg Ca Sr Ba
Li Na K

You will note that this includes some elements yet to be isolated (such as fluorine) but placed according to the behaviour of their compounds. Because of the imprecision of measurement of electrochemical character this list deviates widely at times from a modern order of electrode potentials, though in broad outline it is remarkably sound. Berzelius himself was insistent that his series were approximate only, but nevertheless 'more correct than any other in giving an idea of chemistry'.

In 1819 Berzelius further develops his theme by including now the additional concept of atomic polarization as the proximate cause of electrochemical phenomena. The model he creates is based upon the macroscopic behaviour of tourmaline, which on heating becomes electrically polarized, i.e. opposite charges develop on opposite ends of the crystal. Atoms are deemed to behave in a similar way. As a further refinement, he brings in the concept of unipolarity, supposing that the electricity is more concentrated in one pole than in the other. The atom will then be electronegative or electropositive according to which pole predominates. To explain why, for example, oxygen combines more readily with the predominantly negative sulphur than with the predominantly positive lead, he supposes that the *intensity* of the polarity varies. Although sulphur is predominantly electronegative, its positive pole can still neutralize more electricity in oxygen than can the smaller positive pole in lead. An attempt to interpret this point diagrammatically is given in Figure 1.10, the circles representing the degrees of positive and negative polarization. The important point is that sulphur, though chiefly negative, still has its positive pole with a higher intensity than that of lead.*

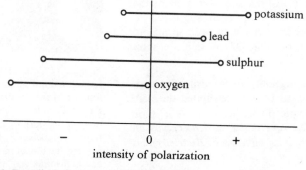

Figure 1.10 Berzelius' views on atomic polarization.

* This paragraph and Figure 1.10 are taken from the author's introduction to a reprint of Berzelius' 1819 *Essai* (Johnson Reprint Corporation, 1972).

Now consider a basic oxide. This would consist of a metal that was electropositive, having a surfeit of positive electricity, combined with oxygen which was always negatively charged. In a similar way sodium chloride would be made of positive sodium and negative chlorine, and on electrolysis these charges would be neutralized and the free elements would be liberated. It is a mistake to think of these 'charges' as our electronic units (the electron was not discovered until 1897). In fact they were not usually units of any kind, and were rarely equal in magnitude. In calcium oxide, for instance, there would be an overall excess of positive charge; similarly in carbonic oxide (CO_2) there would be an excess of negative electricity. Consequently these two oxides can combine, and only in calcium carbonate is anything like overall neutrality attained. As Berzelius said, 'salts are combinations of oxides'; we can indicate it schematically thus:

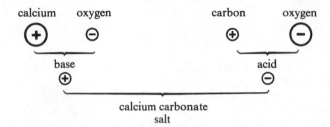

Such a system is essentially *dualistic*. It breaks up all combinations into pairs, ultimately. For Berzelius it was one of the two major keys to chemistry, in which he was followed by a great body of lesser mortals, especially in Germany where chemistry was a thriving science.

SAQ 2 (*Objectives 2, 5*) Using representations like that above for calcium carbonate (but ignoring relative magnitudes of $(+)$ and $(-)$ signs), formulate in Berzelian terms the following:

(a) sulphur dioxide

(b) potassium sulphide

(c) sodium sulphate

(d) potassium hydride

(e) anhydrous alum, $K_2SO_4 \cdot Al_2(SO_4)_3$

The other key principle in Berzelian chemistry was the atomic theory. Again it is not easy to imagine how chemistry had existed without it, but the fact remains that it was not until the first few years of the last century that 'chemical atoms' became available to the theorists. We say 'chemical atoms' because atomism of one sort or another had been in vogue at least since the time of Newton. Often these were conceived as little more than points in space, centres of 'power' or chemical force. But it was left to the Cumberland Quaker John Dalton* (Figure 1.11) to propose, among other things, that each chemical element was composed of atoms, all of which were the same weight as each other but different from those of other elements. Dalton's 'atomic weights' enabled him to make sense of a number of puzzling facts, among these the laws of combining quantities that were just then coming to the forefront of the chemical consciousness. *Why* did eight grams of oxygen always combine with one gram of hydrogen to form water, or with eight grams of sulphur when that was burned? Dalton would say that those gross figures reflected the different weights of the atoms involved. His methods of determining the atomic weights were primitive and arbitrary. Thus he took the atomic weight of oxygen (relative to hydrogen) as 8 in view of the combining quantities. This *assumes* that only one atom of each was involved, and that assumption, for Dalton at least, was another act of faith springing, it seems, from a desire to see nature as simple as possible. Of course it was soon to lead to real trouble, but that is another story.

Berzelius, stimulated by reports of Dalton's ingenuity in far-away England, proceeded to develop atomic ideas of his own. In 1819 he celebrated the marriage between atomic and electrochemical theories in his *Essai sur la Théorie des Proportions Chimiques et sur l'Influence Chimique de l'Électricité* (*Essay on the theory of chemical proportions and the chemical influence of electricity*). So

* On Dalton, see TV programme 2.

Figure 1.11 John Dalton (1766–1844), English natural philosopher and founder of the chemical atomic theory. Born in Cumberland he spent most of his working life as a private tutor in Manchester; he also discovered the law of partial pressures and in his time was a noted meteorologist.

strong was his conviction that this indeed was the key to all chemical progress that even where there were obvious difficulties, as in organic chemistry, Berzelius could assert that electrochemical principles applied here too.

At this point it might be useful to try and identify some of the difficulties that did stand in the path of the electrochemical theory.

SAQ 3 (*Objectives 2, 5, 10*) The following were all facts or theories known in Berzelius' day. Which were compatible with his electrochemical theory and which were not?

1   The displacement of copper by zinc from a solution of copper sulphate.

2   The evolution of heat in many chemical combinations.

3   The combination of two volumes of hydrogen with one volume of oxygen to form water.

4   The existence of diatomic molecules like H_2 and O_2.

5   The reaction we should write as

$$AgNO_3 + KCl = AgCl + KNO_3$$

6   The effect of a direct current on sugar solution.

7   The evolution at the cathode of hydrogen in electrolysis of aqueous solutions of potash.

Thus, to sum up, it looked as though all was poised for the incorporation of organic chemistry into the total structure. That this did begin to happen is due also to a significant new development in analytical techniques—their effective application to organic substances themselves.

1.2.2   The application of analysis: a practical tool

At this point we take our leave, for a while, of details of theoretical schemes and focus attention on a new practical development. Again it happens that the chemist most associated with this advance was Berzelius.

1.2.2.1   Berzelius pioneers a new technique

In 1814 Berzelius was faced with a crisis of confidence. Despite his assured affirmations of 1811 he was seriously wondering whether organic compounds could be considered from the viewpoint of the atomic theory. Assuming that they could he had attempted an analysis of oxalic acid ($C_2H_2O_4$) and obtained a result so absurd that he doubted if it meant anything much at all. In fact he seems to have been dogged with misfortune over this, including a series of experimental errors and a final misprint in his paper. It seemed that one atom of hydrogen was attached to more than 12 other atoms, and the result was quite at variance with previous evidence on inorganic compounds.

After this experience he determined to improve his technique so as to test the opinion that organic compounds could not be viewed atomistically. He was not the first to attempt quantitative organic analysis but he did achieve quite new levels of accuracy. He wrote of the attempt as 'the most difficult I have ever worked on, but I comfort myself with the hope that it is the most important of all my work to date, and perhaps more important than anything I may do in the future. The study of organic nature sheds an entirely unexpected light on the inorganic, and you will see how clear chemistry will become and what a firm foundation its theory will acquire through these studies'. He was not guilty of overstatement.

Let us be clear as to the object of the experiments. It was to determine the relative proportions by weight of all the constituent elements in an organic substance and thus, with the aid of atomic weights, the proportions of atoms present.

How is this done today?

Often by mass spectrometry, but there is still a large amount of elemental analysis done by methods that are fundamentally just the same as those of Berzelius.

See S100, Unit 6

In Figures 1.12 and 1.14 are representations of:

(a) Berzelius' apparatus of 1814.

(b) A modern apparatus working on the same principle.

The questions that follow the next paragraph are designed to bring out the principles from the older equipment and, secondly, to enable you to extend and apply them to current practice.

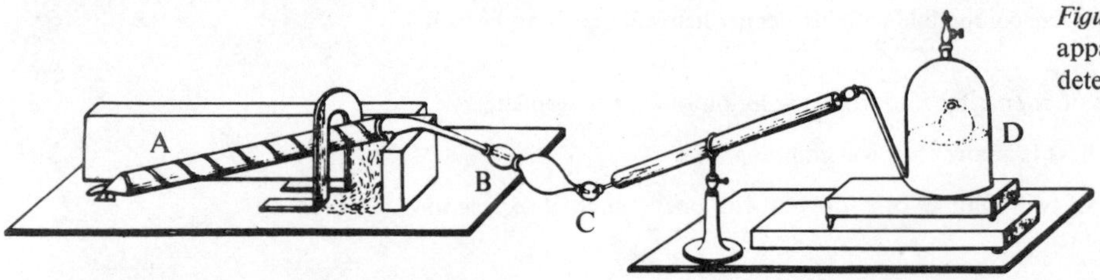

Figure 1.12 Berzelius' combustion apparatus for carbon/hydrogen determinations.

In Figure 1.12, the analysis apparatus A is an iron-clad glass tube, horizontal or sloping, sealed at one end and capable of being heated throughout its length. It contains a known weight of the substance (0.3–0.5 g) with about five times its weight of potassium chlorate mixed with a large amount of sodium chloride to moderate the violence of the reaction. B is an air-cooled bulb and C a tube packed loosely with calcium chloride. D is a bell-jar standing in a trough of mercury, floating on which is a small container with solid potash.

(a) What kind of reaction does the organic compound undergo in A?

(b) What was the purpose of B and C?

(c) What was the purpose of D?

(d) How was the percentage composition calculated from the results?

(e) What difficulties would you expect to be associated with this experiment?

(f) As a final point, how does this diagram compare with modern representations of chemical apparatus?

(a) Oxidation of all the carbon to CO_2 and all the hydrogen to H_2O, most of the oxygen being supplied by the potassium chlorate ($KClO_3$); some oxygen may come from the organic substance itself.

(b) Most of the water would be condensed in B and the rest would be absorbed in C.

(c) All the carbon dioxide would be absorbed by the potash in D.

(d) The increase in weight of B+C is the weight of water formed; as 2/18 of this is hydrogen one can immediately determine its percentage in the original sample. Similarly the increase in weight of D is due to CO_2, of which 12/44 is carbon. In the time of Berzelius and until a few years ago the oxygen was usually determined by difference. A specimen calculation follows on p. 27.

(e) The violence of the original oxidation presented some hazards and it was difficult to weigh the potash vessel without uptake of some atmospheric water and carbon dioxide. One other problem we return to later (p. 28).

(f) The diagram is a rather crude perspective formulation of a kind that survived well into this century but is almost always now replaced by cross-sectional drawings like the one in Figure 1.16 (see p. 28).

SAQ 4 (*Objectives 4, 8*) Which of the following was Berzelius' work on organic analysis designed to test?

(a) The law of conservation of matter: matter can be neither created nor destroyed.

(b) The law of constant composition: each substance will always have constant composition by weight.

(c) The law of multiple proportions: when the same elements combine to form more than one compound, the different weights of one element combining with a fixed weight of the other are in the ratio of small whole numbers.

(d) The belief that life cannot be reduced to chemistry or physics.

(e) The view that Dalton's atomic theory did not apply to organic compounds.

(f) The assertion that animal substances always contained nitrogen.

Several improvements were soon made to this technique by other workers. Potassium chlorate was replaced by copper oxide, CuO (Gay-Lussac, 1815, and many others), U-tubes were introduced for calcium chloride (Bussy, 1822) and Liebig (1830) devised his famous potash bulbs (Figure 1.13) to take the place of the cumbrous apparatus D. These and other improvements devised by Liebig greatly increased the ease and accuracy of organic analysis.

Despite the difficulties Berzelius was able to determine the percentage composition of 13 organic substances in 1814–15. By dividing each figure by the (assumed) atomic weight for the element he was able to arrive at formulae of high accuracy. His work demonstrated that Dalton's atomic theory *was* applicable to organic compounds, a point of great significance as we shall see shortly. Meanwhile it is of interest to note how the experimental data were used. The principle is still applicable today. The following is a typical example, though not due to Berzelius. After combustion of 0.308 g of glucose the calcium chloride absorber increased in weight by 0.188 g and the potash bulbs by 0.448 g. What would be the possible formula for glucose?

Assume atomic weights are $C = 12$, $H = 1$, $O = 16$

$$\text{weight of } CO_2 \text{ produced} = 0.448 \text{ g}$$

$$\therefore \text{ weight of carbon present} = \frac{12}{44} \times 0.448 = 0.122 \text{ g}$$

$$\text{Also, weight of } H_2O \text{ produced} = 0.188 \text{ g}$$

$$\therefore \text{ weight of hydrogen present} = \frac{2}{18} \times 0.188 = 0.0209 \text{ g}$$

$$\text{Hence } \% \text{ carbon in sample} = \frac{0.122}{0.308} \times 100 = 39.7\%$$

$$\text{and } \% \text{ hydrogen} = \frac{0.0209}{0.308} \times 100 = 6.78\%$$

$$\therefore \% \text{ oxygen (by difference)} = 53.52\%$$

Thus the atomic ratios will be obtained by dividing by the masses of the individual particles, i.e. the atomic weights:

C		H		O
$\dfrac{39.7}{12}$	:	$\dfrac{6.78}{1}$	:	$\dfrac{53.52}{16}$

or

3.31	:	6.78	:	3.35

or, approximately,

1	:	2	:	1

Hence possible formulae are CH_2O, $C_2H_4O_2$, $C_3H_6O_3$, etc., i.e. $(CH_2O)_n$. The simplest solution, CH_2O, is the *empirical formula*. To find the *molecular formula*, i.e. the value of *n*, one needs the molecular weight. For solids like glucose this was not determinable until much later in the 19th century. In this case it turned out that $n = 6$ so the molecular formula was $C_6H_{12}O_6$.

You will note that the atomic ratios as determined this way are not exact; they rarely were, owing to experimental errors. Usually C was slightly low and H slightly high. A literal transcription of our results would lead to a nonsense formula like

$$CH_{2.2}O_{1.01}$$

Today it is our faith in the atomic theory that tells us it is nonsense.* But in the time of Berzelius it was the persistent appearance of results that nearly came right that convinced him that behind them all lay a reality that we call 'atoms'.

1.2.2.2 A classical combustion apparatus

A typical combustion assembly as used up until about thirty years ago is shown in Figure 1.14. It will hardly have changed during the previous ninety or so years. The most conspicuous feature is the enormous glass combustion tube nearly 1 m long and heated by up to three dozen separate burners, each individually controlled. Oxygen is passed through for the actual combustion and the gaseous products are then expelled by a slow current of air.

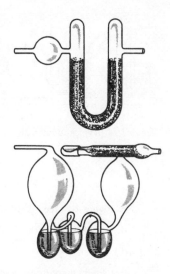

Figure 1.13 Absorption apparatus devised by Liebig.

* By a strange twist of history, those compounds that we recognize today as non-stoichiometric (i.e. without whole number ratios in their formulae) are almost all *inorganic*. It is the *organic* branch that offers the most consistent support for the chemical atomism enquired into by Berzelius.

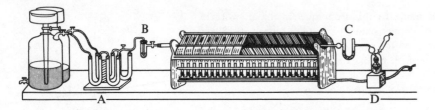

Figure 1.14 Typical combustion apparatus used from mid-19th century.

In Figure 1.14 identify the following:

(i) Bubble counter (to monitor oxygen flow)

(ii) Potash absorber

(iii) Preliminary drying apparatus

(iv) Water absorber

Answer is on p. 31.

1.2.2.3 Some recent developments

There were two serious disadvantages stemming from the scale of the operation. At least 150 mg of a substance were required for each analysis (and often such large quantities were just not available), and a complete run took several hours. In recent years both of these limitations have been largely overcome. The quantities of materials required are now very much smaller, the technique of microanalysis having been pioneered early in this century by Fritz Pregl (who was the first to receive the Nobel Prize awarded for work in analytical chemistry). Moreover, much more sensitive balances are now available, capable of weighing to 10^{-6} g and better.

Figure 1.15 Fritz Pregl (1869–1930), Austrian medical chemist who, at Gratz, revolutionized organic analysis by drastically scaling down the apparatus and modifying the technique. For his development of microanalysis he received the Nobel Prize for chemistry in 1923.

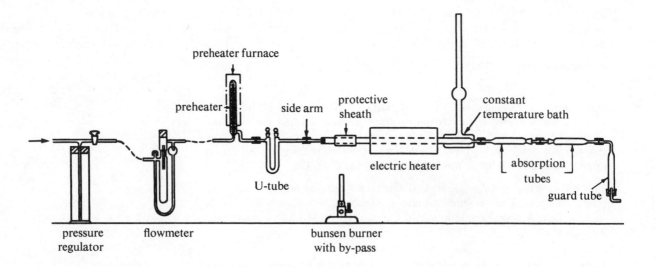

Figure 1.16 Assembly for carbon/hydrogen combustion train, based on British Standard 1428: Part A1 (1958).

Figure 1.16 is a cross-section of a Pregl-type combustion train, as standardized in 1958. Oxygen is admitted at the left-hand end at a rate monitored by the flowmeter. The preheater tube contains platinized asbestos to catalyse the oxidation of any traces of hydrogen in the gas stream; the water formed is then removed by a desiccant in the U-tube. The sample is placed in the combustion tube to the left of the electric heater. Modern practice often favours a rapid combustion method in which the main reaction takes place at 900 °C in a silica combustion tube that is empty though fitted with a series of baffles to retard flow and ensure quantitative oxidation. In this way, combustion is complete in a matter of a few minutes. Other modern innovations include the replacements of calcium chloride by magnesium perchlorate as a drying agent and of potash by soda-asbestos. The constant temperature bath surrounds lead dioxide to remove any oxides of nitrogen, sulphur, etc. More recently, manganese dioxide at room temperature has been used instead.

Flaschenträger tubes (Figure 1.17) are used to prevent contamination by atmospheric moisture or CO_2 when they are removed for weighing; at each end are hollow ground-glass taps which can be closed when combustion is completed. An all-electric furnace assembly is also often used today.

Since the 1960s the method is often automated, with the products of combustion being automatically analysed by gas chromatography. This has made possible further reduction in sample size (to 0.3–3 mg) and in the time of analysis to around fifteen minutes. But the basic principles of combustion are exactly the same.

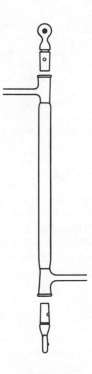

> **SAQ 5** (*Objective 8*) The following results were obtained from a sample of unknown material: 0.015 4 g yielded 0.038 5 g CO_2 and 0.006 8 g H_2O. What was its empirical formula? Only carbon, hydrogen and possibly oxygen were present. As before, C = 12, H = 1, O = 16.

Before leaving this matter of organic elemental analysis it is worth pointing out that a method was soon devised by Dumas (1834) for the determination of nitrogen by combustion. This, again, has survived in modified form to our own day. Figure 1.18 gives a diagrammatic representation of the modern version, though today a mobile electric element is often used to heat the tube, moving gradually along its length from the right.

Figure 1.17 Flaschenträger tube. This is about ¾ actual size. Solid absorbent (e.g. magnesium perchlorate) is kept in place by plugs of glass wool, and the tube may be opened or closed by turning the hollow ground-glass stoppers. (Based on British Standard 1428: Part A5 (1965).)

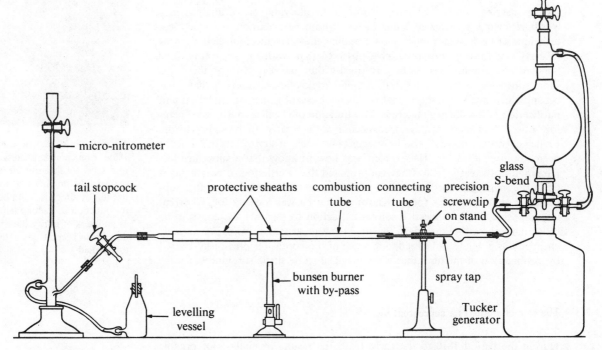

Figure 1.18 Assembly for nitrogen combustion train (micro-Dumas), based on British Standard 1428: Part A2 (1959).

(a) Given that the Tucker generator is simply a source of carbon dioxide, that the combustion tube is filled largely with copper oxide, that the weighed sample is inserted at the right-hand end of the tube and that the micro-nitrometer and levelling vessel are filled with a concentrated potash solution, can you suggest how the apparatus might function?

(b) Copper gauze is usually inserted in the combustion tube towards the left-hand end. Why?

(a) This is a combustion in an atmosphere of CO_2 in which oxygen is supplied by the copper oxide and the nitrogen liberated as N_2 gas. Since CO_2 is absorbed by potash the only gas to survive a passage through that solution, and to emerge into the nitrometer, is nitrogen. Its volume is read off directly and the weight immediately calculated.

(b) In certain cases (e.g. nitro-compounds) some of the nitrogen may be given off as oxides, which are soluble in potash. Accordingly, these are reduced to N_2 by passage over the hot copper gauze.

Finally it is worth noting that other elements can also be estimated in organic compounds (sulphur and halogens at a very early stage). Fairly recently, a useful method for *direct* determination of oxygen has been introduced by Unterzaucher (1940). It is now widely used and involves pyrolysis of the sample in a current of pure nitrogen in the presence of carbon at 1 120 °C. More recently still the use of platinized carbon has permitted reaction at 900 °C. Under these conditions all the oxygen is converted to carbon monoxide, CO. The gases are then passed through iodine pentoxide which reacts quantitatively with carbon monoxide at 120 °C.

$$I_2O_5 + 5CO = 5CO_2 + I_2$$

Estimation of the iodine (by titration) or of the carbon dioxide (by absorption) gives a direct measure of the oxygen in the original sample.

In describing some of these recent developments* we have travelled a long way from the days of Berzelius. Yet for C and H the ideas underlying the techniques have hardly altered.

> SAQ 6 (*Objective 9*) Despite the alternative use of mass-spectrometry in recent times, these 'classical' combustions are still of great importance. Yet it is doubtful if they have ever had the enormous significance that invested them in the early years of the 19th century. Why?

As a postscript to this brief introduction to the unifying role of analysis it is worth reporting a comment from Liebig, whose contributions to analytical technique we have already touched on. Recalling that Berzelius took eight months to carry out thirteen analyses Liebig observed that, with his apparatus, four hundred determinations could be performed within one year. Though there was some substance in the comparison Berzelius took offence, replying that what occupied his time was not the analyses themselves but rather the isolation and purification of the starting materials. This question of chemical purity was therefore raised with new urgency in connection with organic analysis. It was the French chemist, Chevreul, who in his work on the chemistry of fats did most to overcome the difficulties. The problem was how to know that a substance *was* pure, and in the early 1820s Chevreul proposed that a criterion of purity might lie in the melting temperature. This, of course, remains for most organic substances a useful yardstick today. Knowing what he was looking for, Chevreul was now able to apply selective solvent extraction to separate the constituents of his natural starting materials, thus making a great step forward for organic chemistry as a whole and eliminating much of the confusion previously caused by examination of mixtures that were imagined to be single substances.

Figure 1.19 M. E. Chevreul (1786–1889), French chemist notable for his researches on fats and for his remarkably long life—he continued to publish after his 100th birthday. His introduction of melting temperatures as an index of purity took many decades to become generally established.

1.2.3 The role of analogy: a conceptual tool

During the period 1810 to the early 1840s the results of analysis of organic compounds led to the forging of strong links between the old and new branches of chemistry. But there was another factor at work in the same direction, perhaps less obvious but even more powerful and pervasive. This was the growth and development of an approach to organic theory that was rooted and centred in that simplest of devices: an analogy.

* See also TV programme 1.

1.2.3.1 Establishment of the inorganic/organic analogy

Answer to question on p. 28.

(i) B

(ii) D

(iii) A

(iv) C

Consider the situation towards the end of the period. In 1835 a letter to Berzelius from his former student Wöhler conveyed the desperate opinion that 'organic chemistry just now is enough to drive one mad. It gives me the impression of a primeval tropical forest, full of the most remarkable things, a monstrous and boundless thicket, with no way of escape, into which one may well dread to enter'. One may add, with hindsight, that things were to get a lot worse before they began to get better. It is not hard to see why the subject presented such difficulties; an immense number of new compounds had come to light as new techniques were applied to their isolation, purification, analysis and identity. In Germany, especially, the training in laboratory skills of chemists at Giessen, Göttingen and elsewhere simply accelerated the proliferation of all manner of strange novelties. There was an urgent need for theoretical rationalization to enable a path to be hacked through the 'primeval forest', and in fact one had been supplied by Berzelius. The principle that he laid down had its limitations and there was substance in Wöhler's complaint, but without its help the situation would have been far worse; instead of an impenetrable forest an impregnable fortress, perhaps.

Berzelius proposed that 'in exploring the unknown, our only safe plan is to support ourselves upon the known'. This was common sense, not conservatism, but once accepted it led inevitably to the assertion that the only safe guide through the jungle lay in accepting *the analogy between organic and inorganic chemistry*. More precisely, this meant looking at the organic branch in the light of experience in the inorganic, and applying well-tried principles to the new compounds and their reactions. Berzelius maintained this principle to the end of his days, long after Wöhler's outburst and long after other alternatives had been canvassed.

As late as 1846 he wrote:

> The application of that which is already or can hereafter be known concerning the laws of combination in inorganic nature is the only guide to our researches concerning their mode of combination in organic nature; by this means alone we can hope to arrive at a correct and unanimous opinion concerning the constitution of those bodies which occur in nature or which arise from the action of chemical agents upon them.

Much earlier than this, however, he had insisted upon a relationship of analogy, spurred on by his discovery that the laws of atomism applied in both branches. After a lengthy discussion of the laws of chemical proportions in inorganic nature he went on to write (1819):

> The laws which limit the atomic combinations in organic nature are very different from those which we have been examining and permit such a multiplicity of combinations that one can say that no determined proportion exists. The sole phenomenon analogous to these laws that can be discovered therein is that the *substances which have entirely identical properties also have the same composition.** In organic nature the varieties of combination are infinite and have no analogy with those which inorganic nature offers.

One can see the problem. In inorganic nature sulphur, for example, seemed to combine with hydrogen and oxygen in one or two proportions only (in H_2SO_4 and perhaps H_2SO_3, as we should say). But for carbon there appeared to be an inexhaustible number of combinations. So the only analogy perceived at this time was the one we have italicized above. This may not seem very much to us but its first realization did mean a great advance. After all, when one thinks about it for the first time, there is no *a priori* reason why organic substances should obey even this law. In fact, the first flickering glimmer of understanding that they did so was to lead to the establishment of the first fragile tenuous link between organic theory and inorganic theory. As the analogy was worked out in detail so this gossamer-thin connection was to be replaced by other and stouter ones that seemed at the time to be drawing the two branches into strong and permanent union.

* Our italics.

1.2.3.2 General applications of the analogy

There were several advances of great importance where the analogy helped to make sense of what would have otherwise been exceedingly puzzling phenomena. The most notable of these was perhaps the concept of *isomerism*. In the quotation above Berzelius observed that the one common feature of which we can be sure is that substances with identical properties must have identical composition. But was the reverse proposition also true? Did identical composition necessarily imply identical properties? Evidence was beginning to accumulate suggesting it did not.

For example, in 1824 Gay-Lussac and Liebig, working together in Paris, analysed a sample of silver fulminate, while Wöhler, then with Berzelius in Stockholm, obtained the same figures for silver cyanate. Yet the properties were not the same. The fulminate was highly explosive while the cyanate could be heated to its melting temperature. Several acids (lactic, tartaric) were known, of which different samples had different effects on polarized light when this was passed through their solutions. Then, in 1828, came the spectacular transformation by Wöhler of ammonium cyanate into urea, the significance of which lay not in any overthrow of vitalism but rather in the constitutional identity of two such different substances. As Berzelius was later to say, it demonstrated that 'compounds with the same relative number of atoms of the elements can be different by reason of the different way in which the simple atoms are placed relative to one another' (1833). For this remarkable state of affairs, in 1830 he coined the term *isomerism*. This phenomenon, long after Berzelius' time regarded as a peculiarity of organic chemistry, was nevertheless illumined in a strange way by the inorganic analogy. Some years earlier (1819) Mitscherlich had recognized that similarly constituted inorganic salts may have identical crystalline forms (e.g. $ZnSO_4.7H_2O$ and $MgSO_4.7H_2O$) and had proposed the term *isomorphism*. Here lay the clue that Berzelius needed. In writing to Mitscherlich (a former pupil of his) he claimed:

Figure 1.20 J. L. Gay-Lussac (1778–1850), French chemist, discoverer (independently of Dalton and Charles) of 'Charles' law' of expansion of gases by heat, and also of the law of combining volumes of gases. He made important contributions to cyanide and iodine chemistry, volumetric analysis and the manufacture of sulphuric acid.

> You have shown that whenever different elements are combined in the same proportions and in the same manner, identical crystalline compounds are formed. It will now be shown that when the same elements are combined in the same proportions but in a different manner [presumably with the atoms in different relative positions], compounds are formed which have different chemical properties and are different in structure.

Thus it was more than a similarity in name that linked organic isomerism with inorganic isomorphism.

What two kinds of isomerism were involved in these discussions?

The cases of the lactic and tartaric acids are examples of stereoisomerism and that branch of it known as optical isomerism. The only constitutional differences between pairs were in the relative spatial arrangements of the atoms. The other two cases, however, involved differences of structure, with atoms linked to each other in quite different ways.
Thus:

Figure 1.21 E. Mitscherlich (1794–1863), German chemist, pupil and disciple of Berzelius. He is most famous for his recognition of the law of isomorphism (1819). Obtaining the hydrocarbon C_6H_6 by decarboxylation of benzoic acid he gave it the name *Benzin*, or *benzene*.

$$H-O-C \equiv N \rightleftharpoons O=C=N-H \qquad H-O-\overset{+}{N}\equiv\overset{-}{C}$$
cyanic acid   fulminic acid
(tautomeric)

$$\overset{+}{N}H_4\ \overset{-}{O}-C \equiv N \qquad\qquad H_2N-\overset{\overset{O}{\|}}{C}-NH_2$$
ammonium cyanate   urea

Many years later (in 1843) Berzelius was considering the inorganic phenomenon of *allotropy*, the existence of elements in different modifications (diamond and graphite, red and white phosphorus, etc.). Recognizing that carbon in organic combination appeared to behave differently in different compounds he asked whether it might not exist within the combination in several forms related to its different allotropes. Once again the analogy proposed an explanation (though in this case not the right one).

With the benefit of hindsight can you see any merit in this particular extension of the analogy?

In fact, modern studies of diamond and of graphite reveal carbon atoms in, respectively, states of sp^3 and sp^2 hybridization, closely analogous to those in saturated alicyclic and in benzenoid aromatic compounds.

Another chemical idea we owe to Berzelius is *catalysis*. In 1835 he introduced the term as follows:

A new power to produce chemical activity belonging to both inorganic and organic nature ... I shall ... call the catalytic power of substances, and decomposition by means of this power *catalysis* ... Catalytic power means actually that substances are able to awaken affinities which are asleep at this temperature by their mere presence and not by their own affinity.

Despite the rather archaic language we can observe that here again organic and inorganic nature are linked. In fact, in the previous and following paragraphs Berzelius had argued that many biological processes (e.g. the hydrolysis of starch in potatoes by the diastase present) was directly parallel to the recently discovered effects of platinum on many reactions in the gas phase. His analogy still held good and a new explanation based upon it opened up a whole new science of enzyme chemistry. As he said, 'we turned to the study of the chemical processes that occur in living nature with the experience that inorganic nature had provided'.

SAQ 7 (*Objective 3*) Which, if any, are true of the following statements concerning the analogy between organic and inorganic chemistry as conceived by Berzelius?

(a) It was entirely an arbitrary device, in the absence of anything better.

(b) It could be deduced from the laws of atomism.

(c) It became a law in its own right.

(d) It gave highly specific direction to research in organic chemistry.

(e) It was an unproven hypothesis.

(f) It was a useful guiding principle.

1.2.3.3 Development in terms of radicals

These few examples must suffice to show how, in general terms, the use of analogy was enabling Berzelius to bring order and sense to vast areas of newly discovered territory in organic chemistry But, at the same time, there was one specific way in which the analogy was to be developed that affected the nature of chemistry in an even more profound manner. It did not originate with Berzelius but it was invested by him with new authority and its effects remain with us to this day. This was the use of the all-important concept of a *radical*.* Dating back to the late 18th century this idea is familiar to us as a group of atoms present in a molecule and conferring on it certain distinctive properties Moreover, such a group can persist unchanged through a whole series of reactions though the rest of the molecule may be changed beyond recognition. In the following example the ethyl radical (C_2H_5) survives unchanged:

$$C_2H_5OH \xrightarrow{HBr} C_2H_5Br \xrightarrow{KCN} C_2H_5CN \xrightarrow{[H]} C_2H_5CH_2NH_2$$

Putting it in this modern form may help us to see more clearly one vital aspect of any theory based upon the idea of radicals; this is that it must be essentially dualistic, i.e. it divides the molecule into at least two parts (the ethyl and hydroxyl in ethanol, for instance), though this is not to say that further division is impossible (as in the propyl group above in propylamine, when it can be seen that the propyl contains an ethyl radical). This way of mentally carving up a substance was possible even in pre-atomic days. Its first major impetus came from Lavoisier, who regarded inorganic acids as oxides of non-metals and organic acids as oxides of radicals containing hydrogen and carbon (cf. Section 1.1.1).

* Not to be confused with the modern usage, which is an abbreviation for *free radical*.

How could one justify this view of inorganic acids, and how could it have been overthrown?

The view is eminently attractive if one ignores the distinction between a 'mere' solution and a compound between an oxide and water. Consider the case of sulphur; when it burns in air it yields a gas which turns damp litmus paper red. *We* know this results from the following sequence of reactions, but until the theory of ionization was available it made just as good sense to suppose that the SO_2 was the acid as that it was some hypothetical non-isolable H_2SO_3:

$$S \xrightarrow{O_2} SO_2 \xrightarrow{H_2O} H_2SO_3 \rightleftharpoons H^+ + HSO_3^-$$

The simplest way of disproving this would be to isolate acids that did not contain oxygen. This was done early in the 19th century when Davy, having looked long for oxygen in chlorine, was forced to conclude it was not there and that 'muriatic acid' (our hydrochloric acid) was therefore not an oxide but (as we should say) just HCl. Also Gay-Lussac showed that prussic acid was made of hydrogen combined with cyanogen, 'a body which, though compound, acts the part of a simple substance in its combinations with hydrogen and metals.'

Despite these discoveries (which tended to be regarded as exceptional cases) Lavoisier's theory was still in sufficiently good shape for Berzelius to restate it in atomic terminology, to give it a strong electrochemical twist and to extend it to organic compounds. Denying Lavoisier's belief that the essential principle of acidity was oxygen he asserted that 'this resides in the radical of the acids'; if this is electropositive the result is a base, if electronegative an acid (i.e. a basic or acidic oxide). Thus 'it is the radical itself and not the oxygen which determines whether the oxide shall be an acid or a base'.

SAQ 8 (*Objectives 5, 10*) Does the series of 'neutral' salts K_2S, K_2SO_3 and K_2SO_4 fit into Berzelius' scheme or not? Give reasons.

But why limit this notion to just acids and bases?

In 1817 the decisive step was taken of explicitly extending the principle to the organic world:

> The peculiar distinction exists that in inorganic nature all oxidised bodies have a simple radical, whereas on the other hand all organic substances consist of an oxide with compound radicals. In the case of plant substances the radical generally consists of carbon and hydrogen, and in the case of animal substances of carbon, hydrogen and nitrogen.

Two years later he asserted that the dualistic electrochemical view was 'applicable to organic chemistry, and each organic product can be regarded as divisible into oxygen and a compound radical'. At this time only a few odd hydrocarbons and one or two organic chlorine compounds were known as possible exceptions.

Now it must not be supposed from all this that Berzelius dictated how all other chemists were to think (though he might have liked to) nor that he ran the whole of European chemistry single-handed from his remote fastness in Scandinavia. There is no doubt that his influence was enormous, mainly via a prodigious literary output, including a magisterial series of annual reports in which he surveyed the whole chemical scene each year from 1821 to 1848. But there were plenty of other independent thinkers who came to similar conclusions, though as time went on many of them shrank from pursuing the analogy to the extent demanded by Berzelius. We must now see why.

This is certainly not the place to enter into a detailed survey of the complexities of organic chemistry in the 1820s and 1830s. To do so would almost surely elicit a cry of despair like that uttered by Wöhler (see Section 1.2.3.1). It is sufficient to say that in the hands of Wöhler, Dumas, Liebig and many others organic chemistry was to make spectacular progress, and there is no doubt at all that much of their inspiration came from the analogy proposed by Berzelius. Thus, consider the work of Dumas and Boullay in 1827. These workers had been concerned with measurements of vapour densities of ethanol ('alcohol') and a number of its derivatives: ether, ethyl chloride ('hydrochloric ether') and ethyl acetate ('acetic ether'). They concluded that all these substances could be written as double compounds involving ethylene.

Figure 1.22 Berzelian nomenclature in agricultural chemistry: a modern reminder of the persistence of a dualistic way of regarding inorganic compounds (1975).

Suggest, using modern formulae, how this could be so.

Ethanol, C_2H_5OH $= C_2H_4 + H_2O$
Ether, $(C_2H_5)_2O$ $= 2C_2H_4 + H_2O$
Ethyl chloride, C_2H_5Cl $= C_2H_4 + HCl$
Ethyl acetate, $CH_3COOC_2H_5 = C_2H_4 + H_2O + {}^{\prime}CH_2CO{}^{\prime}$ *or* $C_2H_4 + CH_3COOH$

These formulations were largely on the basis of composition measurements, both weights and vapour volumes, rather than of chemical relationships.

What chemical relations do you know connecting ethylene with these substances?

The most important ones are these:

$$C_2H_5OH \underset{+H_2O(b)}{\overset{-H_2O(a)}{\rightleftharpoons}} C_2H_4 \underset{-HCl(d)}{\overset{+HCl(c)}{\rightleftharpoons}} C_2H_5Cl$$

The other products usually involve reactions involving ethanol as an intermediate. The hydration of ethylene (*b*), using concentrated sulphuric acid followed by dilution, was demonstrated at about this time by Hennell, though it was imperfectly understood. The formation of ethylene by the passage through hot tubes of the vapours of ethanol (*a*) or ether had been detected. The elimination of HCl from ethyl chloride (*d*) by the action of alcoholic potash had *not*, and in any case occurs to a very small extent. This is also true of the reverse hydrochlorination of ethylene (*c*).

These formulae raised important questions of 'pre-formation'—e.g. was the ethylene present, pre-formed *as such*, in the molecule? Though of little value today, they represent one of the earliest attempts to connect a series of organic compounds by stressing constitutional relationships. Though the formulae do not conform to Berzelius' demands that they should appear as oxides, they did owe much to his dualistic approach and, in particular, to an alleged analogy with inorganic materials, ammonium salts. Thus, compare this view of ethyl chloride with the conception of ammonium chloride as $(NH_3 + HCl)$. Anticipating Berzelius' indignant reply that, in that case, the 'hydrates', ethanol and ether, should be alkaline, they disarmingly suggested that this would have been so had ethylene been water-soluble! In time Berzelius grudgingly accepted the position, calling ethylene the etherin radical. Meanwhile others, including Liebig, suggested instead that an alternative view would be to consider these substances as containing the radical ethyl. The following, in modern terminology, are representations in terms of ethyl. Uncertainties over atomic and molecular weights led to a doubling of the number of ethyl radicals in the first and third.

Ethanol $2C_2H_5OH = 2C_2H_5 + O + H_2O$
Ether $(C_2H_5)_2O = 2C_2H_5 + O$
Ethyl chloride $2C_2H_5Cl = 2C_2H_5 + Cl_2$

More straightforward than the contentious 'etherin' or 'ethyl' cases was the work of Wöhler and Liebig, in 1832, on 'the radical of benzoic acid' Starting with 'oil of bitter almonds' (i.e. benzaldehyde) they carried out a series of reactions in which it was converted, by turns, to benzoic acid, benzoic anhydride, benzoyl chloride, benzoyl bromide, benzoyl iodide, benzoyl cyanide and benzamide; in modern terms (again) these are: C_6H_5CHO, C_6H_5COOH, $(C_6H_5CO)_2O$, C_6H_5COCl, C_6H_5COBr, C_6H_5COI, C_6H_5COCN and $C_6H_5CONH_2$. (S24– students will have met some of these in a Home Experiment.)

SAQ 9 (*Objectives 6, 11*) What radical is common to all these molecules, and what disadvantage would it have presented to Berzelius?

In fact, this disadvantage proved to be a minor stumbling-block to the great Swedish chemist who hailed the work as 'the beginning of a new day in vegetable chemistry', and even the modest authors identified it as 'a point of light promising an entrance' through the 'dark region of organic nature'.

SAQ 10 (*Objectives 6, 12*) Why should this research be so highly regarded?

Figure 1.23 Justus von Liebig (1803–73), German pioneer of organic chemistry, lifelong friend and collaborator of Wöhler (with whom he worked on benzoyl compounds). He later turned to agricultural chemistry (stressing the importance of mineral fertilizers) and nutrition (inventing 'Liebig's Extract of Meat'). His techniques for analysis brought him fame but, in the long term, he was probably much more important for his work on synthesis and his training of chemists in Giessen. He founded and edited for many years his *Annalen*, one of the great German chemical periodicals, which is still being published.

Throughout the 1830s the radical theory dominated chemical thinking and acted as a major spur to research in organic chemistry. The very fact that there could be more than one 'radical' interpretation (as in the case of ethanol and its derivatives) was a sufficient incentive for further investigation. In 1837 Liebig joined Dumas in Paris, and together they issued a manifesto for chemical radicalism. The great analogy had brought them to this conclusion: 'In inorganic chemistry the radicals are simple; in organic chemistry they are compound—that is the sole difference'.

Such is the air of finality about this declaration that one might well imagine the integration between the two branches to have been virtually complete. Only one thing seemed lacking; no one had yet isolated a radical. The delight of Berzelius may be imagined when, in the early 1840s, came the first report of what seemed a genuine compound radical. This was an evil-smelling product obtained from the action of heat on a mixture of potassium acetate and arsenious oxide: cacodyl.

$$H_3C \diagdown \qquad \diagup CH_3$$
$$As - As$$
$$H_3C \diagup \qquad \diagdown CH_3$$

We know it today as tetramethyldiarsine, and it is obviously no 'free radical' at all. But at that time molecular weights were largely the result of guesswork and the simplest supposition was that the product was $(CH_3)_2As$ or an equivalent formulation. That combination certainly does persist unchanged through many reactions and this seemed a genuine enough radical. According to Berzelius 'this research is a foundation stone of the theory of compound radicals of which cacodyl is the only one whose properties correspond in every particular with those of simple radicals'. Some of the cacodyl reactions are summarized below in modern symbols.

$$CH_3COOK + As_2O_3$$

Figure 1.24 R. W. Bunsen (1811–99), German chemist and pioneer in spectroscopy. At Marburg he did his famous work on cacodyl, but its dangers and general unpleasantness deflected him from further organic research. He worked on combustion and gas analyses, playing some part in inventing the burner that was successfully marketed bearing his name.

This proved to be the first of several 'sightings' of radicals by disciples of Bunsen. Within ten years Kolbe had obtained 'methyl' by the electrolysis of aqueous potassium acetate and Frankland was being introduced in the polite society at Marburg as 'the discoverer of ethyl'—from the action of zinc on ethyl iodide. Alas, 'methyl' was later to prove to be ethane and 'ethyl' its dimer, butane. More significantly, Berzelius was gone (he died in 1848) and with him had disappeared all but a few traces of his famous radical theory, based so soundly on his principle of analogy. Organic chemistry was entering on a new phase of independence and things would never be the same again.

1.3 Secession: Moves towards an independent organic chemistry—PHASE III

It is never wise to chop up the past into 'periods', as though history were discontinuous, but around the early 1840s the relationship between organic and inorganic chemistry undeniably entered a new phase. At about this time organic chemistry had developed its own distinctive concepts in sufficient detail for these to become differentiated from those of the parent subject. With this growth went the other changes associated with increasing independence: the establishment of specialist vocabulary, the practice of specialist skills and techniques, even the development of a new ethos or 'atmosphere' and the first signs of a new professional awareness. All this did not happen at once but the signs are unmistakable that, by the mid-1840s, organic chemistry was rapidly developing in new directions that owed much less than before to inorganic precedent.

> **SAQ 11** (*Objective 13*) In 1843 W. H. Perkin spoke of organic chemistry as being 'still in its infancy.' What could he have meant?

This is not to say that the dream of a unified chemistry was allowed to fade. As an ideal it remained, and some of the most ardent advocates of the 'new' organic chemistry were vociferous in their assertions that chemistry was one.

Indeed one is tempted to suppose that the insistence of their claims might have been provoked by a general feeling that, in practice, it was not. And this is really the point. Whatever the chemical theoreticians might proclaim, either because they genuinely believed it or because they saw it as a good public relations exercise, in actual fact organic chemistry was becoming more and more divorced from the older branch.

Figure 1.25 The Royal College of Chemistry, London.

An interesting example is provided by A. W. Hofmann, who from 1845 to 1865 was Professor of Chemistry at the newly founded Royal College of Chemistry in London. In 1853 he addressed the Royal Institution at the beginning of a course on organic chemistry. Referring to many attempts to arrive at 'a rigorous distinction between inorganic and organic compounds' he concluded that 'the separation of chemical science into inorganic and organic is by no means found in nature'. This denial could be—and was—defended by excellent theoretical arguments; it can also be seen as an adroit political move at a time when a desperately needed Government support for science was more likely to be forthcoming if science could present a united front. Only a few months before, Lyon Playfair had been urging for the same reason a unity between 'practical' and 'theoretical' men of science. And Hofmann was speaking at the very time when negotiations for Government support for the RCC were most hectic. By October that year the support had been granted.

In practice Hofmann, of all people, was a living demonstration of the separation of organic chemistry. Of the 360 research papers he produced only the merest handful were *not* in that area, and it has been argued that it was he, more than anyone else, who set the fashion for organic research in England. He must have been aware of a slight incongruity in his position for he wryly admitted to his audience, 'I am almost afraid, gentlemen, that you will object to me that in denying the distinction of inorganic and organic compounds, I lose the very ground upon which I stand'.

What Hofmann's audience thought is not recorded, but few of them, at the end of the course, could have been unaware of the spectacular new directions organic chemistry was then taking. There seem to have been at least three discernible factors behind the new developments and we shall look at these in turn, noting especially their effect on the organic/inorganic interface.

Figure 1.26 A. W. Hofmann (1818–92), German organic chemist who directed the Royal College of Chemistry in London from 1845 until his appointment as Professor in Berlin twenty years later. His research centred on aromatics derivable from coal-tar, and he is remembered for the Hofmann degradation of amides to primary amines and the methylation technique for amines, which also bears his name.

1.3.1 The 'failure' of electrochemistry

Throughout the 1820s and most of the 1830s chemistry had been dominated by the electrochemical theory of Berzelius. Wielding his principle of analogy between the two branches, Berzelius had ensured that organic chemistry was in no way exempt from that domination. The unification of chemistry in that sense might have reached an astonishing degree of completeness had it not been for the gradual build-up of awkward facts which even the ingenious Berzelius was hard-pressed to fit into his dualistic scheme. In the end the most telling arguments against his analogy came from the discovery of the simple fact of organic substitution.

It is hard to realize how little was known about organic reactivity 150 years ago. A great mass of curious information about natural products was rapidly accumulating and some of the degradation reactions of what we now recognize as very complex molecules were being studied. But perusal of the literature of the time reveals a puzzling difference from modern work that is not at first easy to identify. What is largely missing is the study of *simple* reactions on *simple* molecules. The reason, of course, is that it was not at all obvious then what were simple substances or reactions. But with hindsight we can spot the omission. This may help to explain why an apparently simple reaction like halogen substitution came to light in an unplanned, almost casual, manner, and, when it did, was regarded for a long while as something of an oddity.

Credit for making the first systematic assault on this problem must rest with the French chemist Dumas, who was introduced into the subject in an unexpected and bizarre manner. At a Government soirée in the Tuileries every effort had been made to gain a high standard of excellence, even to the extent of using candles of specially bleached wax. The consternation can be imagined when these splendid-looking candles began to give off clouds of choking fumes and proceedings were brought to a speedy and embarrassing conclusion. Dumas was called in to investigate the unfortunate incident and soon established that the candles had been bleached in chlorine and that the choking gas was hydrogen chloride; consequently some of the chlorine must have entered the wax and chemically combined with it. This led him to a general investigation of the action of chlorine on organic compounds and, in particular, its ability to replace some of the hydrogen. Among other reactions he examined was the rather complex action of chlorine on ethanol to give chloral and the alkaline decomposition of this to yield chloroform (see Unit 9):

Figure 1.27 J. B. A. Dumas (1800–84), French chemist, later achieving high office in government. His studies of aliphatic chlorination led to his theory of substitution and the decline of electrochemical dualism. He devised a useful method for vapour density determination and another for the estimation of nitrogen in organic compounds.

He did not, of course, express it in these terms, and, although he was able to deduce correctly the formulae of chloral and chloroform, the reaction sequence was far too complicated to suggest an obvious incompatibility with Berzelian dualism. He also examined the action of chlorine on turpentine; this is a highly unsaturated substance and addition must have taken place, but what did seem striking was that some hydrogen was removed as HCl and for every hydrogen atom that was ejected one of chlorine took its place. After much further study he concluded (in 1834) that 'chlorine possesses the singular power of separating the hydrogen from certain bodies and of replacing it atom for atom'—a special case of a 'law of substitutions'.

SAQ 12 (*Objectives 5, 10*) This discovery was to prove an acute embarrassment to Berzelius. Why?

At first Berzelius' indignation was muted. Then, in 1838, Dumas announced the discovery of trichloracetic acid from the prolonged chlorination of acetic acid itself, emphasizing the similarity between the two. This time the effect on Berzelius and on all organic theory was far more serious. To understand why we must first look at Berzelius' formulation of acetic acid.

Given that the empirical formula of acetic acid is CH_2O, that its molecular weight is 60, that Berzelius (for reasons that do not matter here) doubled the correct molecular weight, and that he regarded such compounds as analogous to sulphuric acid, which he wrote as

$$SO^3 + H^2O,$$

write down his formulation of acetic acid. ($C = 12$, $O = 16$, $H = 1$)

If we take the compound radical as analogous to a sulphur atom, $C^4H^8O^4$ would become $C^4H^6O^3 + H^2O$ or, more fully, $C^4H^6 + 3O + H^2O$.

The obvious formula for the chlorinated product would thus be

$$C^4Cl^6 + 3O + H^2O$$

but this would be to admit direct replacement of electropositive hydrogen by electronegative chlorine, a complete violation of the cherished dualistic principle. As Berzelius wrote to Wöhler, this kind of formula 'necessarily involves the overthrow of the whole structure of chemistry in its present form, and the revolution is based on the decomposition of acetic acid by chlorine'. Accordingly he took refuge in an alternative arrangement for the product. This was a result of 'reshuffling' the 16 atoms then thought to be present in an acetic acid molecule. Two substances already known were oxalic acid (whose hypothetical anhydride was C^2O^3) and 'perchloride of carbon', C^2Cl^6 (hexachloroethane, discovered by Faraday from prolonged chlorination of ethylene). With those 'ingredients' Berzelius rearranged his formula for acetic acid as follows, the C^2Cl^6 part being called a *copula* (meaning a kind of connecting link); it might have seemed a master-stroke to borrow this word from his arch-enemy Gerhardt and to invest it with quite a new meaning, but it was later to raise his foes to more passionate fury than ever. He now wrote:

$$H^2O + C^2O^3 + C^2Cl^6$$

Even as early as this, formulae were regarded as expressing chemical behaviour in some sense, and the gross disparity between this formula and the one advocated for acetic acid lent an air of unreality to the whole exercise. After all, the two acids were similar in many ways. Table 4 compares some of their properties:

TABLE 4

	acetic acid	trichloracetic acid
appearance	colourless liquid	colourless crystals
melting temperature	16.6 °C	52 °C
boiling temperature	118.2 °C	195 °C
solubility in water	complete	very soluble
action of alcohol	esterifies	esterifies
decarboxylation	heat with soda-lime	boiling water
pK_a	4.76	0.66

In today's terms what principle would seem to be contravened by such a proposal as that of Berzelius?

Today, with the theory of structure firmly established (see Unit 2), we would find difficulty in accepting that similar substances might differ greatly in structure. However, there are plenty of exceptions; for example, sugar and saccharin, though both sweet, are totally dissimilar in constitution. A more serious objection lies in the principle that structure tends to be conserved through a reaction; compare the preservation of the ethyl structure in the sequence of reactions on p. 33. We must not read our ideas into pre-structual chemistry, of course, but there does seem to have been an intuitive feeling towards some such principle even then.

In passing we can note (possibly in Berzelius' defence?) that we now recognize many cases where basic structures do break down, sometimes unexpectedly. A famous example is the rearrangement of 'neo-pentyl' to 'iso-pentyl' compounds during S_N1 reactions but not during S_N2 reactions:

The first reaction thus involves the breaking and remaking elsewhere of one carbon–carbon bond. Recently the following example has been reported with no less than 20 such processes involved (something of a record and for amusement only!):

(K. V. Scherer (1972) *Tetrahedron Letters*, 2077)

This difficulty was intensified by another discovery that followed swiftly on the heels of Dumas' original one. In 1844 his assistant Melsens announced that the chlorinated acid could be easily reduced back to acetic acid, this making quite untenable the view that they had unrelated kinds of formulae. By this time Berzelius was in serious difficulty and could only resort to changing the acetic acid formula to bring it into line with the new one for its chlorinated product. Accordingly he rewrote acetic acid as $H^2O+C^2O^3+C^2H^6$; thus we have:

$$H^2O+C^2O^3+C^2H^6 \underset{\text{reduction}}{\overset{\text{chlorination}}{\rightleftharpoons}} H^2O+C^2O^3+C^2Cl^6$$

acetic acid   (tri)chloracetic acid

The similarity is now represented but Berzelius is forced to concede substitution of hydrogen by chlorine if not in a radical at least in a *copula* (whatever that was). He was bitterly assailed by his opponents, who were by now disenchanted with other aspects of the dualistic scheme. As one of his most virulent critics, the French chemist, Laurent, was to write, 'What then is a copula? A copula is an imaginary body, the presence of which disguises all the chemical properties of the compounds with which it is united'.* Unable or unwilling to see the rationale behind Berzelius' position he went on to protest: 'the dishonesty is flagrant'.

Certainly from now on (the early 1840s) dualism went into such a decline that the whole inorganic/organic analogy could no longer survive in the simple form proposed by Berzelius. On a short view it certainly looks as though he was greatly in error. However, the longer view that we can enjoy may not only modify our judgement somewhat, but may also offer a clue as to why things seemed to go so badly wrong. With this in mind we shall digress for a short while to glance at the modern theory of aliphatic chlorination.

Figure 1.28 A. Laurent (1807 or 1808 to 1853), French chemist who, despite a failure to obtain a satisfactory academic post, was a brilliant organic theoretician who contributed much to the assault on dualism. He worked extensively in the naphthalene field and collaborated closely with Gerhardt; it has been suggested, not unfairly, that his greatest service was in being the one person capable of interpreting Gerhardt's work.

* Presumably because it was supposed not to exert an appreciable influence on the rest of the molecule. Hence its appearance in a formula would 'disguise' the real behaviour of the compound.

It has to be admitted that the chlorination of aliphatic carboxylic acids at the α-position is not a simple process. Often traces of phosphorus are used as catalyst and, under those conditions, it is believed that the acyl chloride (e.g. CH_3COCl) is the reacting species. However, there is strong evidence that chlorination *can* occur in the same kind of way as for alkanes (paraffins). Certainly in Dumas' experiment the indications are that this was the case. He allowed reaction to occur in strong sunlight, a condition necessary for the symmetrical fission of a chlorine molecule (homolysis) into two chlorine atoms, each highly reactive owing to its unpaired electron. A chain reaction then takes place as follows. The organic compound is represented simply as CH_3R and a dot represents an unpaired electron, a double dot (:) indicating an electron pair.

$$Cl:Cl \; \underset{}{\overset{light}{\rightleftharpoons}} \; Cl\cdot \; + \; Cl\cdot$$

Then,

$$Cl\cdot \; + \; H:\overset{\overset{\textstyle H}{\cdot\cdot}}{\underset{\underset{\textstyle H}{\cdot\cdot}}{C}}:R \; \longrightarrow \; Cl:H \; + \; \cdot\overset{\overset{\textstyle H}{\cdot\cdot}}{\underset{\underset{\textstyle H}{\cdot\cdot}}{C}}:R$$

Then,

$$Cl:Cl \; + \; \cdot\overset{\overset{\textstyle H}{\cdot\cdot}}{\underset{\underset{\textstyle H}{\cdot\cdot}}{C}}:R \; \longrightarrow \; Cl\cdot \; + \; Cl:\overset{\overset{\textstyle H}{\cdot\cdot}}{\underset{\underset{\textstyle H}{\cdot\cdot}}{C}}:R$$

followed by attack by the $Cl\cdot$ atom on another molecule, etc.

If this free radical mechanism is accepted, then it can be seen that the initial attack of the chlorine atom was not on the carbon but on the hydrogen atom; in other words, the first step involving the organic compound is abstraction of hydrogen to form hydrogen chloride, and this is entirely consistent with the Berzelian view. The next step is then the attack of the newly formed organic radical on the only species present in any large quantity, a chlorine molecule. It is here that the earlier theory was deficient, in refusing to admit a diatomic molecule of the type Cl_2. Its real weakness lay not in proposing a polarity in molecules, but in insisting that this was present in all of them all the time. To suggest, as some historians have done, that dualism is applicable to electrovalent compounds characteristic of inorganic chemistry but not to the usually covalent organic counterparts, is to miss a most important chemical point. We recognize today that there are degrees of polarization as exemplified in the following series:

$$Na^+Cl^- \quad H{-}Cl \quad CH_3\overset{\overset{\textstyle O}{\|}}{C}{-}Cl \quad (CH_3)_3C{-}Cl \quad CH_3CH_2{-}Cl \quad Cl{-}Cl$$

$$\xrightarrow{\text{————— decreasing polarization —————}}$$

but, except in the totally symmetrical molecule Cl_2, there will be at least a small electrical polarity, just as Berzelius maintained. He even proposed that molecules could have temporary polarizations as external conditions changed; in this again he was far ahead of his time. Had the concept been applied to the diatomic chlorine molecule, with the polarization towards Cl^- and Cl^+ given consideration, the difficulty of H^+ being replaced by chlorine, as Cl^+, would have vanished. But for Berzelius this was unacceptable. It may be remarked that the concept of positive halogen is of use today in accounting for some electrophilic additions to alkenes and substitutions in the aromatic field, though usually the 'positive halogen' is in fact associated with solvent molecules.

SAQ 13 (*Objectives 3, 4, 5*) What was the fundamental weakness of the electro-chemical theory that led to its rejection in the 1840s?

There were other areas in which the theory ran into difficulties, notably in the theory of acids. Laurent's younger colleague, Gerhardt, proved that acetic and chloracetic acids had formulae half those assigned to them by Berzelius, i.e. $C_2H_4O_2$ and $C_2HCl_3O_2$, the latter proving that H_2O could not be present and

Figure 1.29 C. F. Gerhardt (1816–56), French chemist whose short career was plagued by ill-health, dearth of research facilities, personal animosities and general insecurity. Despite these difficulties, Gerhardt, together with Laurent, succeeded in undermining much of the dualistic hold over chemistry and replacing this, at least temporarily, by the theory of types. His attempts to revise atomic and molecular weights were of real importance.

that the analogy with H_2SO_4 (SO_3+H_2O) could not hold: 'there is no water in our acids and no oxide in our salts' (1843). But the rock on which the analogy really foundered was the fact of substitution. It was ironic that this happened just at the time when the hypothetical radicals were being produced as triumphal vindications of the theory. Unfortunately for Berzelius they arrived just too late to save the situation.

For a quarter of a century his analogy had provided not only a guide to progress in organic chemistry, but, it can be argued, the only guide that was possible. Where else could have been found a directive principle into such unknown territory? Now, its usefulness outworn, it was to be replaced, or, perhaps one should say, turned on its head.

1.3.2 The inversion of the analogy

With the threatened demise of dualistic organic chemistry the leadership of chemical thought began to slip away from Berzelius and strenuous efforts were made to assert it in France. It was a French chemist, Dumas, who in 1835 perceived a new aspect of the unity of all chemistry. He wrote:

> The future progress of general chemistry will be due to the application of the laws observed in organic chemistry ... and far from limiting myself to taking the rules of mineral chemistry to carry into organic chemistry, I think that one day, and perhaps soon, organic chemistry will give rules to mineral chemistry.

This was a remarkable anticipation of what was, in fact, to become a theme of Parisian chemistry within a few years. No one was more clearly aware of what was happening then than Berzelius who, in a passionate exchange of letters with Laurent in 1843–4, laid before him the following serious charge:

> You are endeavouring to reform the theory of inorganic chemistry according to ideas you have derived from your experience with organic compounds. In my chemical studies I have preferred the diametrically opposite route, that of basing speculations about organic composition on the theoretical ideas more or less established for inorganic compounds.

Just as Dumas had predicted, the analogy was now being used in the opposite direction. This reversal had several interesting consequences. One was that it gave new life to a belief that had been held intermittently for many years, namely that our 'elements' may not be as simple as we suppose. Davy had raised this question and now the Irish chemist, Robert Kane, proposed that 'simple' (inorganic) radicals may be of the same kind as 'compound' (organic) radicals. Arguing *from the organic experience* he suggested that in one case we know the composition, in the other we do not. Very much in the same spirit Laurent and Gerhardt were to point to known complex radicals in *inorganic* chemistry, such as ammonium or phosphate, while Dumas was to extend his law of substisution to cover replacement of other elements than hydrogen—even carbon or metals; he noted that the isomorphism of $KMnO_4$ and $KClO_4$ implied that Mn might be replaced by Cl.

This last proposal called forth a satirical letter to Liebig's *Annalen*, signed by 'S. C. H. Windler' (German for swindler) but actually written by Wöhler. It purported to describe what happened when all the atoms in manganous acetate were replaced, stepwise, by chlorine. The result was, of course, pure chlorine, though it retained all the original properties of the manganous acetate. Bleached fabrics consisting of this 'spun chlorine' were on sale in London and 'much sought after'! Behind the heavy humour was a serious chemical point. Dumas was indeed proposing that in substitutions (like chlorination) the properties of reagent and product (as $KMnO_4$ and $KClO_4$) do not necessarily differ very much and Wöhler was trying to reduce that argument to absurdity. According to Dumas 'there exist certain types which are conserved' through such reactions. This theory of 'types' was in fundamental contrast to the radical theory in that it tended to look at molecules whole, rather than in terms of hypothetical fragments. Modified by both Laurent and Gerhardt it came to be styled a 'unitary theory' for that reason. Its fortunes do not need to detain us here. It is sufficient to point out that, in Gerhardt's hands, it came to involve a classification of organic compounds into four 'types', based on ammonia, water, hydrogen chloride and hydrogen (1853). Thus the water type could include alcohols, ethers and acids, e.g.

$$O\begin{cases}H\\H\end{cases} \quad O\begin{cases}C_2H_5\\H\end{cases} \quad O\begin{cases}C_2H_5\\C_2H_5\end{cases} \quad O\begin{cases}C_7H_5O\\H\end{cases}$$

water   alcohol   ether   acid

These formulae did not represent how the atoms were arranged in the molecule, merely how they appeared to react. Nor did they represent a break with the inorganic tradition; four inorganic substances were the basis of the whole scheme, and it may be noted that Laurent's early work was guided by another inorganic analogy, the concepts of crystallography dictating his interpretation of many curious facts associated with the chemistry of naphthalene. And the continued emphasis on the unity of all chemistry ensured that inorganic chemistry was not forgotten. But its principal generalization was now no longer the guide to the organic chemists, and their newly won independence was marked by several important conceptual advances. Laurent seems to have been the first to arrive at the simple conclusion that an organic substance is just one containing carbon, while Gerhardt in effect acknowledged the unique ability of carbon to form long chains in his concept of homologous series, families of related organic compounds (e.g. alcohols) where each member differed from the one above it and the one below by the group CH_2 (1843).

> **SAQ 14** (*Objective 11*) In the light of this what is particularly remarkable about the following assertion of 1847 (from the 8th edition of Turner's *Elements of Chemistry*, edited by Liebig and Gregory)?
>
> As the organic world is characterised by the predominance, in quantity, of carbon, so the mineral or inorganic world is marked by a similar predominance of silicon, and by the very great abundance of chlorine (in sea-salt), and of the metals sodium, aluminium, calcium and iron. Oxygen, hydrogen and nitrogen are very abundant in all the kingdoms of nature.
>
> It is thus evident that organic chemistry is not essentially different from inorganic chemistry, so far as the nature of the elements is concerned. But there is a peculiarity in the mode in which the chief elements of organic nature are combined together; and it is this which we express by the term 'Chemistry of Compound Radicals'.

If, as it appears, one may detect in this quotation the strong influence—even if not the actual words—of Liebig, one may be reminded that his position was rather different from that of the French school. But in its own way it was quite as influential in achieving a high degree of autonomy for organic chemistry. Liebig, more than anyone else, was the driving-force behind the early stages of the movement that we now turn to consider.

1.3.3   The rise of synthetic organic chemistry

Of the two factors in the movement which we have described in terms of a greater independence for organic chemistry one was the failure of an analogy and the other was its inversion. The third factor was very different, being almost sociological in nature. This was the establishment of a strong tradition of synthetic organic chemistry in Germany, a development that was bound up with the emergence of Germany as a nation, the gradual professionalization of science and (above all) the beginnings of sound and systematic laboratory instruction.

In 1825 Liebig became Professor at Giessen.* For 28 years he supervised the chemical laboratory that was soon to acquire world fame for its course of practical instruction and its research output. This pioneer of chemical education was fortunate in persuading the State of Hesse-Darmstadt to give funds and he made the most of his privilege. Students came from all over Europe to discover the joys of experimental chemistry and to learn its special skills. The

---

* See TV programme 1.

Figure 1.30 Liebig's analytical laboratory at Giessen.

whole subject was taught, but an emphasis on organic chemistry research was associated with Liebig's own interest in perfecting the techniques of organic analysis. By 1830 he was in a position to apply them with confidence to the immense numbers of new compounds isolated or prepared. When, in 1832, he launched his own chemical journal the *Annalen* (still published today and known as *Liebig's Annalen*), he acquired a further instrument with which to extend his influence (and, it may be added, to belabour his foes). Partly as a result, organic chemistry, much more than the inorganic branch, became the hallmark of German chemical studies by the 1840s. It was also forging strong links with the pharmaceutical industry. Other centres arose, and Marburg, Heidelberg, Göttingen, Munich, Bonn and Berlin eventually became the envy of European chemists, all with a strong organic emphasis. During this period any Englishman who wanted a research training would almost certainly spend some time at one of the German universities, with perhaps a stay at Paris also. There was nothing for him at home until Hofmann imported the German methods into the Royal College of Chemistry in 1845.

Some indication of the enormous influence of Liebig may perhaps be gained from the diagram below, which indicates the names of some of his students together with the places with which they came to be most closely associated:

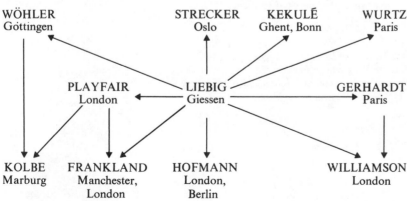

WÖHLER
Göttingen

STRECKER
Oslo

KEKULÉ
Ghent, Bonn

WURTZ
Paris

PLAYFAIR
London

LIEBIG
Giessen

GERHARDT
Paris

KOLBE
Marburg

FRANKLAND
Manchester,
London

HOFMANN
London,
Berlin

WILLIAMSON
London

It is undoubtedly the case that the expansion of organic chemistry as promoted by Liebig and his heirs and successors helped materially to establish its autonomy during the 1840s and 1850s. This can be discerned in two rather different respects, theoretical and practical.

In matters of organic theory Liebig by no means went the whole way with Dumas, Laurent or Gerhardt. German chemistry was much more reluctant to abandon the dualistic views of Berzelius, and it was not until 1858 that the French type theory was introduced into Germany (by Limpricht at Göttingen). In fact, however, Liebig had convinced himself by 1839 that 'Berzelius fights for a lost cause' and was concluding that his principle of analogy was strictly limited: 'Up to a point therefore we follow the principles of inorganic chemistry, but beyond the point where they let us down we need new principles'. Thus instead of inverting that analogy he was severely restricting it, so breaking from *both* extreme wings of chemical theory. This was a declaration of independence of which neither would have approved, but its implications for the unity of chemistry were crystal-clear. From now on at least some of the organic chemists were to go their own way.

This was not to mean the establishment of a 'neutralist' or 'moderate' faction of chemical theory, equally zealous for their cause. German organic chemistry for a long time became far less committed to all detailed theoretical schemes than its French counterpart. Thus in 1841 Liebig unfavourably compared the French propensity to wonder whether radicals were types with the German dedication to productive experiment in the laboratory. Years later he observed 'with the theory of substitution as a foundation organic chemistry needs only labourers'—not clever theoreticians.

The second divergent tendency in German organic chemistry was in the development of its own characteristic techniques. (See TV programme 1.) This was an area where superb manipulative technique was essential. Given the opportunity to learn it, the German research worker would soon find himself confronted with problems of isolation and purification that constituted enough of a challenge in their own right. Moreover, as any organic chemist today will confirm, the sheer satisfaction of producing white crystals from the most unpromising black tar or gum is so great that it can easily become almost an end in itself. Chemists would vie with each other in displays of experimental expertise, and to that end acquired all manner of useful skills.

The experience of Edward Frankland was not untypical. Working in Bunsen's laboratory at Marburg, he became the first Englishman to graduate there as a Ph.D. During this time he acquired considerable skill in glass-blowing and became the first chemist to develop the technique of heating in a sealed tube, i.e. under pressure, later passing on the techniques to Hofmann. He also pioneered the use of a Liebig condenser in the reflux position and the employment of ethyl iodide as a laboratory reagent. In ways like this the organic chemists lived and worked in a world that in practice was very different from that inhabited by their inorganic colleagues. And so, of course, were the problems that they had to face: enormous numbers of new compounds, problems of isomerism, polymerism and much else then thought to be unique to their subject. Then, in the 1860s came totally new and unexpected happenings that gave to organic chemistry a perspective that changed its very nature and once again brought the two branches together. These developments are so important for modern chemistry that the whole of the next Unit is devoted to them.

SAQ 15 (*Objectives 2, 13, 15*) In 1816 Sir Humphry Davy complained: 'the substitution of analogy for fact is the bane of chemical philosophy; the legitimate use of analogy is to connect facts and to guide to new experiments.' How far would that have been a fair comment on the general development of chemistry to 1860? Suggest an outline for a possible essay on this subject.

Time chart: Some important dates

1789	Lavoisier's *Traité*
1800	Electrolysis discovered by Nicholson and Carlisle
1801	
1802	
1803	
1804	
1805	
1806	Davy's work on electrolysis begins
1807	
1808	Dalton's *A New System of Chemical Philosophy*
1809	
1810	
1811	Berzelius' electrochemical theory and Avogadro's hypothesis announced
1812	
1813	
1814	Berzelius' analysis of organic compounds
1815	
1816	
1817	
1818	
1819	Berzelius' *Essai*; discovery of isomorphism
1820	
1821	
1822	Research on 'etherin' by Dumas and Boullay
1823	
1824	
1825	
1826	
1827	
1828	Wöhler's 'synthesis' of urea
1829	
1830	Isomerism recognized by Berzelius
1831	
1832	Researches on the benzoyl radical (Wöhler and Liebig)
1833	
1834	Dumas' theory of chlorination
1835	Catalysis recognized by Berzelius
1836	
1837	Liebig and Dumas' theory of organic chemistry
1838	Chlorination of acetic acid (Dumas)
1839	
1840	
1841	
1842	Reduction of trichloracetic acid (Melsens)
1843	
1844	
1845	
1846	
1847	
1848	Death of Berzelius
1849	
1850	
1851	
1852	
1853	Gerhardt's type theory

Time chart: The rise and fall of electrochemical dualism

A highly schematic summary; this must not be taken too literally as there is no vertical scale: how does one measure 'importance' of a discovery? It simply indicates the main steps without identifying the height of each.

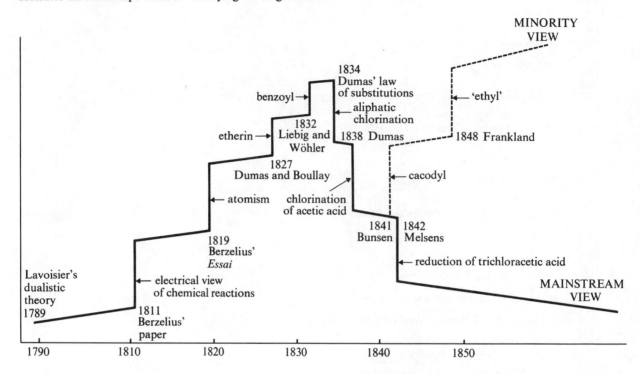

SAQ answers and comments

SAQ 1 The numbers refer to Table 2.

(a) No: benzene was not yet discovered.

(b) The answer is 'No' but you may have got it right for the wrong reasons; though ethylene does not appear in the list the impure gas was known (from ethanol)—hence the dichloride (no. 24). The dibromide could not have been formed since bromine was not discovered till 1826 (Table 1).

(c) Yes; ethanol (5) and benzoic acid (9) were both known, as was sulphuric acid for a catalyst.

(d) No; magnesium was not isolated until 1808 (Table 1). The Grignard reagents were discovered in 1900.

(e) Yes; it had already been obtained by oxidation of sucrose (1) with nitric acid (Scheele, 1776).

SAQ 2

(a)

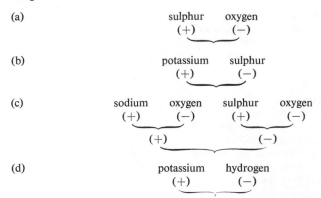

(b)

(c)

(d)

(e) This is an example of what Berzelius called a third-order salt—made of two second-order salts joined together. Such compounds would be held together by very feeble 'degrees of affinity', i.e. the residual charges would be minute:

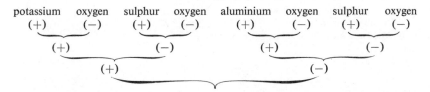

There would be an even smaller residual charge at the end capable of uniting the molecule to (negative) water of crystallization.

If you still find difficulty with (say) sulphur being (+) in some compounds and (−) in others, read again the paragraph preceding Figure 1.10 on p. 23.

SAQ 3 1 This is entirely consistent with Berzelius' scheme since it included an electrochemical series with zinc well above copper in electropositive character.

2 If a reaction does involve a neutralization of electric charge, heat might be expected as in a flash of lightning. He wrote 'In every chemical combination there occurs a neutralization of opposite electricities, and this neutralization produces fire'.

3 This fact is neither for nor against the theory. Avogadro interpreted it to mean that two atoms of hydrogen would combine with one of oxygen (since on his hypothesis equal volumes of all gases contained equal numbers of 'molecules'). Berzelius accepted this deduction, writing water as H^2O (or later as H_2O) but it did not bear directly on his theory. See also Unit 2, Section 2.2.

4 This view was incompatible with the electrochemical theory. If all molecules were dipolar, with a positive and a negative part, the parts had to be different, not the same.

One could have HCl but not H_2. Hence Berzelius interpreted the hydrogen/oxygen reaction as

$$2H + O = H_2O$$

and not
$$2H_2 + O_2 = 2H_2O$$

(You may have noted that this means in effect accepting Avogadro's hypothesis for the reactants but not the product, since the volume of steam produced equals that of the hydrogen consumed.)

5 This presented no problem for 'the electric polarization will... be better neutralized in the latter combination than in the former'.

6 Another serious problem, highlighting the differences between organic and inorganic substances. There was no way of explaining why one should be an electrolyte and the other should not, for there is, of course, no effect whatever in the case of sugar solution.

7 The discharge of hydrogen instead of potassium is consistent with Berzelius' scheme as with modern ionization potential theory. The less electropositive element is discharged first.

SAQ 4 (c) and (e).

SAQ 5 $C_7H_6O_2$.

SAQ 6 Because they showed that atomism applied to organic as well as inorganic substances and that, from that viewpoint at least, chemistry was *one* subject. It was at least worth considering that guidelines already established in inorganic chemistry may also be applicable in the new subject of organic compounds.

SAQ 7 Of all these suggestions only (f) seems unequivocally correct. Although there was 'nothing better' it certainly was not *arbitrary* (a) because it was suggested by the discovery that atomism reigned over both territories. But one could not possibly say it followed *logically* from atomic laws (b) in the sense that it was not derived from them directly. It was neither a law (c) nor a hypothesis (e), and its direction was general rather than specific (d). It has been termed a 'regulative principle'—something that regulates our approach to science and scientific theories but not one of these theories itself.

SAQ 8 The characterization of K_2S as a salt is *not* consistent with Berzelius' scheme because he held (p. 34) that salts are combinations of basic and acidic oxides. However, the existence of the sulphite and sulphate are compatible, and go some way to confirming the view that oxygen is not the essential principle of acidity. If it were, SO_3 and SO_2 would not *both* be able to neutralize K_2O. Of course we now know that, because of the weakness of sulphurous acid, potassium sulphite has a more alkaline reaction, and therefore in a sense is 'less neutral' than potassium sulphate.

SAQ 9 A first glance may suggest C_6H_5, but inspection shows the radical is obviously C_6H_5CO, termed by Liebig and Wöhler 'benzoyl'. Its difficulty for Berzelius was that he had proposed that oxygen should be outside, not inside, a radical.

SAQ 10 For Berzelius and his followers it was clearly a vindication of his belief in organic radicals. In more general terms, it was among the earliest sustained pieces of evidence that such radicals could persist unchanged through many reactions; it removed the fear that 'etherin' or 'ethyl' might be special cases.

SAQ 11 Possibly the fact that it had only existed as a separate science for a mere two or three decades in comparison with the centuries of chemistry itself. But more probably Perkin was speaking of emergence *in the 1840s* of a relatively independent discipline with its own characteristic theory.

SAQ 12 Because an electronegative element, chlorine, was replacing an electropositive one, hydrogen. If the organic compound before substitution were to be conceived as a dualistic combination of hydrogen (+) and the rest (−), then how could chlorine (−) be attached to the rest?

SAQ 13 We can see that the electrochemical dualistic approach to organic chemistry was rejected, not because it was inherently inappropriate here and the famous analogy a false one but because in matters of detail there simply was not evidence to enable it to be modified in the right direction.

SAQ 14 Nowhere in this statement is there any sign of recognition that carbon is *responsible* for the difference.

SAQ 15 This raises some fairly basic questions as to how science in general proceeds. Among the important points which should be made are the following:

1 *Analogy of organic with inorganic chemistry*
Various aspects of this analogy proposed by Berzelius and others include points of advantage and disadvantage. On the credit side the analogy did serve to 'connect facts' as in the formulation of organic acids, in the discovery of catalysis, and the interpretations of isomorphism and allotropy. On the debit side the reformulation of trichloracetic acid could well be said to involve replacement of analogy for facts, and this became increasingly obvious as more information about substitution became available.

2 *Analogy of inorganic with organic chemistry*
Notions about the complexity of the elements and of inorganic substitution were fair deductions from the analogy, though at the time it was hard to see the experimental basis.

3 *Other analogical points*
Many other cases of analogy have been described, some of which led to genuine advances in research even though, by today's standards, the analogies were false. One may mention the oxygen theory of acids, the behaviour of tourmaline on heating and, especially, the analogies within organic chemistry itself which led to the postulation of similar formulae for substances showing similar behaviour and the eventual emergence of the idea of homologous series. Furthermore, the development of the theory of types itself was a superb example of analogical reasoning.

In these suggestions we have not come out on one side of the argument or the other as this is a matter on which you can legitimately be expected to form your own opinion. But that opinion should certainly take into account some of the facts mentioned above.

Further reading

The following suggestions for further reading are strictly optional but may be of some interest to those who wish to explore some of the issues in more depth than has been possible in the Unit.

(a) *Books*
John Read (1957) *Through Alchemy to Chemistry*, Bell, London, chapters 9 and 10.

A. Findlay and T. I. Williams (1965) *A Hundred Years of Chemistry*, Methuen, London, chapters 1 and 2.

A. J. Ihde (1966) *The Development of Modern Chemistry*, Harper and Row, New York, Evanston and London, chapters 4–8.

J. E. Jorpes (1966) *Jac. Berzelius, his Life and Work*, trans. B. Steele, Almqvist & Wiksell, Stockholm.

T. H. Levere (1971) *Affinity and Matter*, Clarendon Press, Oxford, chapters 5 and 6.

C. A. Russell (1972) Introduction, Commentary and Notes to the Johnson Reprint edition of Berzelius' *Essai sur la Théorie des Proportions Chimiques et sur l'Influence Chimique de l'Électricité* of 1819, Johnson Reprint Corporation of America.

(b) *Papers*

J. H. Brooke (1968) Wöhler's urea, and its vital force—a verdict from the chemists, *Ambix*, **15**, 84–114.

J. H. Brooke (1971) Organic synthesis and the unification of chemistry, *Brit. J. Hist. Sci.*, **5**, 363–92.

J. H. Brooke (1973) Chlorine substitution and the future of organic chemistry, *Stud. Hist. Phil. Sci.*, **4**, 47–94.

C. A. Russell (1963) The electrochemical theory of Berzelius—I: origins of the theory, *Ann. Sci.*, **19**, 118–26.

C. A. Russell (1963) The electrochemical theory of Berzelius—II: an electrochemical view of matter, *Ann. Sci.*, **19**, 127–45.

C. A. Russell (1968) Berzelius and the development of the atomic theory, in D. S. L. Cardwell (ed.) *John Dalton and the Progress of Science*, Manchester University Press, pp. 259–73.

Acknowledgements

Grateful acknowledgement is made to the following for material used in this Unit:

Figures 1.1–1.4, 1.7–1.9, 1.11, 1.19, 1.23, 1.26 and 1.27 From S. Muspratt (1853) *Chemistry, theoretical, practical and analytical as applied and relating to the arts and manufactures*, Vol. 1; *Figure 1.5* From A. Lavoisier (1789) *Traité élémentaire de chimie*; *Figures 1.6 and 1.29* Chemical Society Library Portrait Collection; *Figure 1.10 and paragraph preceding this figure* C. A. Russell and Johnson Reprint Corporation of America, from C. A. Russell's introduction to the reprint edition (1972) of the 1819 *Essai sur la Théorie des Proportions Chimiques* by Berzelius; *Figure 1.15* From *Les Prix Nobel 1923*; *Figure 1.16* From BS 1428: Part A1, 1958, British Standards Institution; *Figure 1.17* From BS 1428: Part A5, 1969, British Standards Institution; *Figure 1.18* From BS 1428: Part A2, 1959, British Standards Institution; *Figure 1.21* Royal College of Physicians; *Figure 1.22* Fisons Ltd.; *Figure 1.24* Picture Archive, Philipps University, Marburg; *Figure 1.25* Imperial College of Science and Technology; *Figure 1.28* Günther Bugge (1929) *Das Buch der grossen Chemiker*, Verlag Chemie Gmbh, Weinheim/Bergrstr; *Figure 1.30* An engraving by Trautschold. c. 1840. *Cacodyl reactions on p. 37* From C. A. Russell (1971) *The History of Valency*, Leicester University Press.

Unit 2 Synthesis and structure

Contents

2.0 Introduction

2.0.1 The state of the argument

In Unit 1 we have briefly followed the changing shape of chemical science from, very roughly, 1800 to the late 1850s. Obviously we have been highly selective in the events we have looked at, the criterion of choice being how far they affected the relationship between organic and inorganic chemistry. And this, may we remind you once again, has been to enable us to have a better apprehension of the state of chemistry today. By the end of Unit 3 you should begin to see clearly why we have been looking at the past quite so intently. Before proceeding to examine the chemical ideas that came to prominence in the 1860s we shall offer a quick résumé of the changes recorded in the previous Unit. The progress of the relatively new science of organic chemistry was seen to have been marked by three phases of development in relation to its older relative, inorganic chemistry.

Phase I: Recognition

Although some kind of distinction between mineral and vegetable substances had been recognized since the late 17th century, it was around 1800 that a study of materials from living organisms accelerated to such a rate that a 'new science of organic chemistry' was capable of being clearly recognized. Once this had been done (and the name has, of course, stuck ever since) there was an obvious tendency for the subject to develop along its own distinctive lines. So, in the early 1800s, we have the first definite signs of a divergence between the two chemistries, the new and the old, the organic and the inorganic (as it became known). However, almost as soon as this happened efforts were made to establish whether this divergence was desirable or even necessary. These led to a further phase in their relationship.

Phase II: Incorporation

Organic chemistry emerged as a rapidly growing body of knowledge just at the same time as chemistry was being infected with the new chemical atomism of John Dalton. Many people did in fact come to regard this as something tantamount to a disease, but that was later. At the time its tendency to spread and permeate chemistry to its furthest bounds suggests the medical metaphor. The question then was: is organic chemistry going to be immune? It soon became apparent that it was not, and that atomism applied just as much in the new area as in the old. This followed from the effective manipulation of a new practical tool, the technique of organic quantitative analysis. And still, today, essentially the same techniques corroborate the same conclusion. But given this new convergence, with natural products conforming to the laws of atomic combination, how could their very different compositions and behaviour be understood? For Berzelius, and many after him, this was to be made possible by the exploitation of an analogy, by which organic chemistry was to be interpreted in terms of the facts and theories of the inorganic branch. This proved to be a conceptual tool of immense power, and to many in the 1830s it must have appeared that the ultimate union of the two branches was just round the corner. This, however, was not to be.

Phase III: Secession

Just when the triumph of Berzelius' analogy seemed about to be complete the whole structure of organic theory collapsed about his head—or so it must have seemed. Fatally linked with a dualistic view of organic compounds the analogy found difficulty in surviving the wholesale rejection of that approach by the French chemists, who, instead, stood it on its head and thenceforth interpreted inorganic in terms of organic chemistry. This in itself, of course, was a tribute to the enormous progress made in the new science under the old regime; it was now able to dominate its former parent! A Declaration of Independence does not automatically mean a breaking of old ties; merely a shift in the balance of power. Had events been confined to France this is possibly all that would have happened. But in Germany and elsewhere the rise of synthetic organic chemistry led to such an exclusive concentration in that area that for practical purposes its growth was independent of the older branch. Unification remained a lofty ideal, to be sure, but the reality was very different (as Unit 3 will show).

That, then, was the position in the 1850s. In the present Unit we take up the story with the injection into chemistry of new concepts that were to change it utterly and to cause the two divergent branches to swerve together again in a union that was to seem as permanent as it was illusory. Loud were the cries of 'chemistry is one' and, in a sense, they were entirely justified. But behind the shouting and the propaganda we can detect a genuine note of agreement as organic and inorganic chemists for a time made common cause in response to the new stimuli upon them. In a word these stimuli may be summarized as those of *synthesis* and *structure*. Emerging from the first, and leading to the second, was the fundamental doctrine of *valency*. This in its turn was largely dependent upon the clarification of *atomic weights*, a process that owed much to an international chemical conference at Karlsruhe.

This Unit will discuss all those matters and will be chiefly concerned to show how events in the 1860s led to a real convergence between organic and inorganic chemistry. From time to time, however, we shall have to go back to the period covered by Unit 1 in order to point up the significance of events between 1860 and the 1870s. One important thing to emerge in that period is a general clarification of chemical formulae, and this was to be as important for chemistry as the convergence of its two branches. As we shall show, it helped to contribute towards that convergence. But it was primarily synthesis and structure that led to the convergence of the two branches of chemistry. It is interesting that modern demands for a unified chemistry (Unit 1, Section 1.0.4) on the basis of synthesis, structure and dynamics were anticipated in the first two respects by over one hundred years.

As you read this Unit you should therefore have a twofold general aim: to understand why the 1860s constituted a watershed for chemistry in respect of (a) the meaning of chemical formulae and (b) the unification of chemical science. More precise objectives are given below.

2.0.2 Objectives

1 To identify the characteristics of the approach to synthesis of Berthelot, Perkin and Frankland.
(SAQ 1)

2 To evaluate the effectiveness of Berthelot's syntheses as (a) important preparative methods or (b) powerful agencies in reforming chemical theory.
(SAQ 5)

3 To evaluate the significance of Frankland's discovery of diethylzinc for chemistry generally.
(SAQs 2, 3)

4 To recognize the different chemical meanings of the term 'synthesis'.
(SAQ 4)

5 To calculate atomic weights on the bases of (a) Berzelius' guidelines and (b) modern ideas, including Avogadro's hypothesis.
(SAQs 6, 7, 9)

6 To evaluate the importance for chemistry of Cannizzaro's revival of Avogadro's hypothesis.
(SAQ 9)

7 To distinguish between the separate elements in the concept of valency as it emerged in the period 1852–64.
(SAQ 10)

8 To identify the main contributions to the theory of valency of Kekulé, Frankland, Couper, Crum Brown and Williamson.
(SAQ 10)

9 To characterize the essential features of the theory of structure.
(SAQ 11)

10 To recognize and evaluate structural representations by Kekulé, Crum Brown and Frankland.
(SAQs 12, 13)

11 To evaluate the roles of (a) valency and (b) structure in the unification of chemistry in the 1860s.
(SAQ 14)

12 To recognize the limitations of planar representations of three-dimensional molecules.
(SAQ 11)

13 To identify the factors causing chemistry to show greater unification in the 1860s than in the period before.
(SAQs 8, 14)

Table A

List of scientific terms, concepts and principles in Unit 2

Introduced in a previous Unit	Unit Section No.	Developed in this Unit	Page No.
	S100[1]		
Avogadro's hypothesis	5.3.3	atomic heat	21
chirality	10.4.5	catenation	30
dissociation	9.12	saturation capacity	28
free radical	13.2.1	structure, theory of	33
tetrahedral carbon atom	10.4		
valency	10.3		
	S24–[2]		
enantiomorphism (optical isomerism)	10.4.5		
organometallic compounds	7.2.2		
synthesis	9.1		
	S25–[3]		
molecular orbitals	8.3		
	S351[4]		
isomorphism	1.2.3.2		
vital force	1.2.1.3		
	T & N[5]		
dimerization	p.65		
resonance	p.91		

[1]The Open University (1971) S100 *Science: A Foundation Course*, The Open University Press.

[2]The Open University (1973) S24– *An Introduction to the Chemistry of Carbon Compounds*, The Open University Press.

[3]The Open University (1973) S25– *Structure, Bonding and the Periodic Law*, The Open University Press.

[4]See Unit 1.

[5]J. M. Tedder and A. Nechvatal (1966) *Basic Organic Chemistry*, Vol. 1, Wiley.

2.0.3 Related Course material

TV Programme 2 This programme is about the use of models in chemistry and begins by some location film in places in the Lake District associated with the early years of John Dalton. From Dalton's models we pass to a consideration of numerous other kinds of atomic models, including those of Hofmann and Kekulé, concluding with the tetrahedral models used by van't Hoff. Our main purpose in the programme is to show how a choice of the right kind of model not only reflected chemical thinking at the time, but also helped to determine it in the future. The culmination of this programme, with van't Hoff's models of the early 1870s, is intended as a bridge leading into the next three programmes, which are all related to stereochemistry.

Radio 1 See also Unit 1.

2.1 Organic synthesis

2.1.1 The claims of Berthelot

We have already encountered organic synthesis in connection with the growth of synthetic organic chemistry in Germany (Unit 1, Section 1.3.3). We shall meet it again in Unit 3 and, indeed, throughout the Course. One of the aims of this Section is to explore the multiplicity of meanings associated with this phrase as well as to show how deeply the concept has penetrated into the philosophy of chemistry. So far, we have noted that synthetic advances, however loosely we define them, had tended to insert a wedge between the two main branches of the science. It may therefore come as something of a surprise to read the following extract from a book of 1860. This was *Chimie Organique fondée sur la Synthèse* (*Organic Chemistry based upon Synthesis*) by the French chemist Marcellin Berthelot.

> Only synthesis can establish in a definitive manner the identity of the forces which act in mineral chemistry with those which act in organic chemistry, by showing that the former are sufficient for producing all the effects and all the compounds to which the latter give rise.

As if this were not clear enough a few pages later he asserts that the operation of organic syntheses 'in principle obliterates all the demarcation lines between mineral and organic chemistry'.

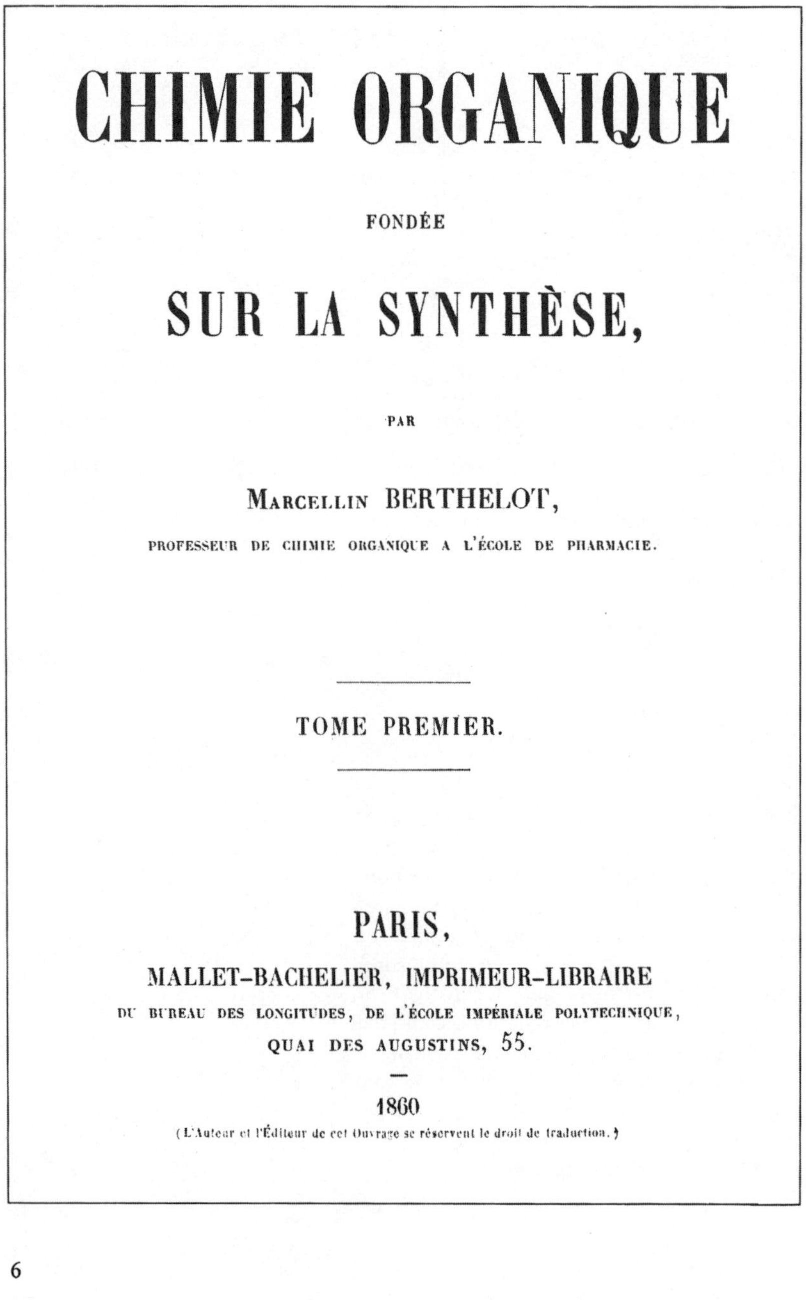

CHIMIE ORGANIQUE

FONDÉE

SUR LA SYNTHÈSE,

PAR

MARCELLIN BERTHELOT,

PROFESSEUR DE CHIMIE ORGANIQUE A L'ÉCOLE DE PHARMACIE.

TOME PREMIER.

PARIS,

MALLET–BACHELIER, IMPRIMEUR–LIBRAIRE

DU BUREAU DES LONGITUDES, DE L'ÉCOLE IMPÉRIALE POLYTECHNIQUE,

QUAI DES AUGUSTINS, 55.

—

1860

(L'Auteur et l'Éditeur de cet Ouvrage se réservent le droit de traduction.)

Figure 2.1 Title page of Berthelot's *Chimie Organique fondée sur la Synthèse* (1860).

In the light of Unit 1 what is remarkable about this claim?

Two things seem particularly remarkable:

1 Wöhler's synthesis of urea (1828) might have been expected to have produced this effect, but, as we have seen (Unit 1, Section 1.2.1.3) this was not the case although it has been widely believed to be so.

2 The practical effect of organic synthesis was partly to deepen, not obliterate, the 'demarcation lines' separating it from inorganic chemistry.

It is clear that Berthelot believed that the production of organic compounds by synthesis would prove 'the identity of forces' in inorganic and organic chemistry by elimination of the 'vital force' from the latter. Yet, as Dr John Brooke has so clearly shown,* the convergence between the two branches that had undoubtedly taken place by the 1860s owed little to the Wöhler 'synthesis'. In his view the use of analogical reasoning was chiefly responsible for those tendencies to unite that did exist. The contention of these Units is that superimposed upon that effect is the more complex one of synthesis. This operated at two levels, practical and ideological. At first the practical differences in techniques inevitably pushed the two branches further apart—as in German chemistry in the 1840s and 1850s. Now, however, Berthelot was claiming that synthesis would *in principle* overthrow the distinction between the forces at work in the two divisions of chemistry. To appreciate the force of his argument we must examine his own synthetic work more closely.

Taking issue with Lavoisier's description of chemistry as 'the science of analysis', Berthelot asserted that this was wrong because it was incomplete; analysis was only half the chemist's job. A better approach would be this:

> Having completed his analytical work the chemist proposes to re-compose what he has destroyed; for his point of departure he takes the ultimate degree of analysis, i.e. the simple bodies. He compels them to unite with each other and, by their combination, to re-form the same natural principles which constitute all material beings. Such is the object of chemical synthesis.

For organic chemists the first target would be the hydrocarbons from which one may obtain alcohols, and then aldehydes, acids, esters, etc. At an early stage Berthelot discovered two methods for preparing methane, CH_4. One was the action of potash at high temperatures on carbon monoxide:

$$CO + KOH = HCOOK,$$

followed by pyrolysis of the formate produced; another was the copper-catalysed interaction between hydrogen sulphide and carbon disulphide at high temperatures:

$$2H_2S + CS_2 = CH_4 + 4S$$

Although the yields were small the starting materials could all be obtained directly from their elements. This enabled him to complete his 'circle of metamorphoses' in which ordinary laboratory-based reactions could ensure not only the breakdown of organic compounds but also their synthesis. One such 'circle' is given below (modern formulae):

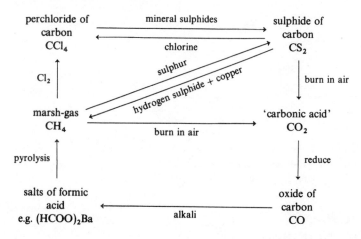

Ambix, 1968, **15**, 84–114; *Brit. J. Hist. Sci.*, 1971, **5**, 363–92.

Another cluster of synthetic reactions centred around the hydrocarbon acetylene, C_2H_2. In 1862 Berthelot showed that this gas could be produced by passing an electric spark between carbon electrodes in an atmosphere of hydrogen:

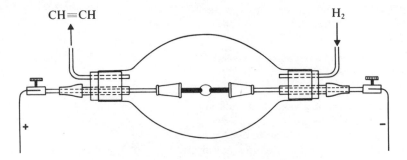

$CH{\equiv}CH$ H_2

Figure 2.2 Berthelot's apparatus for the production of acetylene.

Both in technique and in ideology this bore a remarkable resemblance to the much later work by Miller (1953) seeking for evidence for the 'primeval soup' theory of the origin of life.

Reduction of acetylene was shown to give ethylene and ethane, sparking in a mixture of acetylene and nitrogen led to production of hydrogen cyanide (prussic acid), and passage of acetylene through a red-hot tube led to the production of numerous polymeric materials, including benzene. Under similar conditions benzene formed diphenyl, mixtures of benzene and acetylene led to naphthalene, and naphthalene and acetylene led to acenaphthene. These reactions are given in modern terminology below:

$$C_2H_2+H_2 = C_2H_4 \qquad C_2H_4+H_2 = C_2H_6$$
$$C_2H_2+N_2 = 2HCN$$

A final illustration of Berthelot's work must come from his studies on the formation of alcohols from hydrocarbons. In 1854 he dissolved ethylene in concentrated sulphuric acid (to form ethyl hydrogen sulphate) and decomposed this with water to produce ethanol:

$$C_2H_4+H_2SO_4 = C_2H_5HSO_4$$
$$C_2H_5HSO_4+H_2O = C_2H_5OH+H_2SO_4$$

This had been done before by Hennell but it was not well-known. Berthelot also obtained propanol from propene, and showed that a general method existed in the prior conversion of an alkene to the halide by addition of hydrogen halide, followed by hydrolysis to the alcohol.

In these and other ways Berthelot pursued for ten years the quest for general methods for organic synthesis. His motivation was clearly stated:

> To banish life from all explanations relating to organic chemistry, — that is the aim of our studies. Only in this way shall we succeed in constituting a science that is complete and self-consistent.

Figure 2.3 Marcellin Berthelot (1827–1907), French chemist noted for a determined attack on the problem of organic syntheses and for important contributions to the new science of physical chemistry, especially in his studies of heats and velocities of reactions. He wrote extensively in the history of chemistry, especially that of the ancient world. Like Dumas he held high office in the French government, becoming Foreign Minister in 1895.

Berthelot's strongly anti-vitalist tendencies have been sometimes seen as an expression of a deeply felt religious agnosticism, and (although there is no necessary logical connection between the two) it does seem likely that this was true in his case. In a similar way his scepticism extended to a disbelief in atoms themselves. But whatever may have been the motivation for such an intense decade of work there can be little doubt that any lingering traces of chemical vitalism were finally dispelled by these organic syntheses.

SAQ 1 (*Objective 1*) In what respects did Berthelot's synthesis differ from that of Wöhler?

2.1.2 Some other contenders

When Wöhler read Berthelot's magnum opus on synthesis he wryly observed that it might convey the impression that scientific organic chemistry and chemical syntheses originated with the author. This was simply not true and we must now glance back to the earlier period. In 1858 the English chemist Edward Frankland delivered a lecture at the Royal Institution 'On the production of organic bodies without the agency of vitality'. During this he referred to two earlier pieces of work which, he said, 'completely broke down the barrier between the so-called 'organic' and 'in-organic' bodies'. These were Wöhler's urea synthesis (a rare claim by a 19th century chemist) and some interesting research by his friend Hermann Kolbe. This centred around what we should call tetrachloroethylene, C_2Cl_4, obtained by passing carbon tetrachloride vapour through a hot tube. Kolbe showed that on treatment with chlorine and water it yielded (tri)chloracetic acid; this, as we have seen, can be reduced to acetic acid, and therefore the sequence of reactions *could* be said to constitute a synthesis of acetic acid. As Kolbe commented:

> It is an interesting fact that acetic acid, hitherto known only as an oxidation product of organic matter, can be almost immediately composed by synthesis from its elements.

The complete sequence would then be:

$$C+2S \xrightarrow{\text{heat}} CS_2 \xrightarrow{Cl_2} CCl_4 \xrightarrow{\text{heat}} C_2Cl_4 \xrightarrow{Cl_2+H_2O}$$
$$CCl_3COOH \xrightarrow{[H]} CH_3COOH$$

It has been pointed out that this seems to be the first reference to the word 'synthesis' in a paper on organic chemistry. Frankland was not the only chemist to be convinced of the unifying power of this and related work by Kolbe. In an address to the British Association in 1864 William Odling condemned the use of different symbols in the two branches of chemistry, concluding:

> Now that organic and mineral chemistry are properly regarded as forming one continuous whole, a conclusion to which in my opinion Kolbe's researches on sulphuretted organic bodies have largely contributed, it is high time that such relics of the ancient superstition that organic and mineral chemistry were essentially different from one another should be done away with.

It is true there are not many other examples of successful syntheses before Berthelot, but that does not mean that these were regarded as impossible. A classic case was that of the young W. H. Perkin. In 1849 Hofmann had commented on the desirability of artificially producing the substance quinine. Prompted by this remark Perkin set out to try by the oxidation of allyltoluidine with chromic acid. He was unsuccessful, but obtained instead the first synthetic dye, aniline purple, by replacing the allyltoluidine in this reaction by aniline. Synthesis of the drug was prevented, not by lack of motivation, but by sheer lack of understanding of the chemistry involved.

Perkin wrote:

> I began to think how it might be accomplished, and was led by the then popular additive and subtractive method to the idea that it might be formed from toluidine by first adding to its composition C_3H_4 by substituting allyl for hydrogen, thus forming allyltoluidine, and then removing two hydrogen atoms and adding 2 atoms of oxygen, thus
>
> $$2(C_{10}H_{13}N)+3[O] = C_{20}H_{24}N_2O_2+H_2O$$
> allyltoluidine   quinine

What was wrong with his reasoning?

Assuming the starting point was *o*-toluidine, the structures are:

The conversion would be wildly improbable because of the lack of structural relationship between the starting-material and product. One benzene ring would have to be destroyed and three new rings would have to be obtained by cyclization. The point, of course, is that Perkin had no structural formulae and had no theory of structure. It is a good illustration of a theme we shall encounter frequently: the limitation of molecular formulae without a theory of structure.

Figure 2.4 Hermann Kolbe (1818–84), the German chemist who became the successor to Bunsen at Marburg (1851) and then to Liebig at Leipzig (1865). He did important work in organic syntheses, including those of acetic acid (with Frankland), numerous hydrocarbons by electrolysis of carboxylic acid salts, and salicylic acid by carbonation of phenol (the last two methods now bearing his name). As editor of *Journal für praktische Chemie* he was able to express uninhibitedly his anti-structural agnosticism, as in his well-known attack on van't Hoff.

Figure 2.5 W. H. Perkin (1848–1907), the English chemist who prepared (by accident) the first synthetic dyestuff, mauve. His subsequent adventures as a dyestuff manufacturer at Greenford led to an early fortune and retirement. The Perkin reaction which he developed is that between aldehydes and carboxylic acid anhydrides in the presence of the sodium salt of the acid as catalyst. In this way he synthesized coumarin (the odiferous principle of new-mown hay and much in demand in perfumery, etc.).

A further most important development in organic synthesis emerged quite by accident in the work of Edward Frankland. This young Lancashire chemist had spent three formative months during 1847 in the laboratory of Bunsen at Marburg. Inspired by Bunsen's researches on cacodyl Frankland acquired a new and compelling objective:

> The isolation of the alcohol [alkyl] radicals was, at this time, the dream of many chemists, whilst others doubted or even denied their very existence. I was also smitten with the fever and determined to try my hand at the solution of the problem.

Figure 2.6 The building at the University of Marburg which formerly housed Bunsen's laboratory.

As we mentioned earlier Frankland's efforts appeared to be successful since it was commonly believed that he had managed to isolate 'ethyl' (Unit 1, Section 1.2.3.3). In fact he had not, but his work was far more important for quite different reasons. A brief account of this will offer a good insight into the way that chemists tended to think in the middle of the last century.

Frankland, together with his friend Kolbe, decided to try to remove the ethyl radical from ethyl cyanide by means of potassium metal, known to have a strong 'affinity' for cyanide. A reaction certainly took place but it did not give ethyl. An extremely complex sequence of reactions seems to have occurred, including a base-catalysed trimerization of ethyl cyanide to an aminopyrimidine:

Other products included ethane, which Frankland identified as 'methyl'.

Shortly after this Frankland removed to England to take up teaching duties at Queenwood College in Hampshire—one of the earliest schools in England to teach experimental science. With Frankland in charge of chemistry and Tyndall looking after physics it must have been a lively place. When time permitted he tried the effect of various other reagents on ethyl cyanide, but to no avail. He then made the epoch-making decision to attempt the removal of iodine from ethyl iodide, which he heated with potassium and succeeded in obtaining several gaseous products.

Figure 2.7 Queenwood College, Hampshire, where Edward Frankland taught.

What would you expect these to be?

The action of potassium on an alkyl iodide would be to remove the iodine and form an alkyl free radical. This might be expected to undergo dimerization or disproportionation.

For example, in modern notation,

$$C_2H_5I + K\cdot \longrightarrow C_2H_5\cdot + KI$$

Then
$$2C_2H_5\cdot \longrightarrow C_4H_{10}$$

or
$$2C_2H_5\cdot \longrightarrow C_2H_4 + C_2H_6$$

In fact, Frankland obtained ethylene and ethane in equal volumes but did not identify butane ('ethyl'). Altogether the reaction was far too vigorous. Accordingly on 18 July 1848 he tried the much less reactive metal zinc in the hope that this would react in the desired way with ethyl iodide.

Figure 2.8 Edward Frankland (1825–99), the English chemist who played an important part in the recognition of valency and the foundation of organometallic chemistry. He became Professor at Manchester (1851), St Bartholemew's Hospital, London (1857), the Royal Institution (1863) and the Royal School of Mines, successor to the Royal College of Chemistry (1865). His later work was in the area of public water supplies and purification and, in 1877, he became the first President of the (Royal) Institute of Chemistry.

This reaction was carried out, like many of the others, in a sealed tube. His technique was to open these tubes at one end under water and allow gaseous products to collect in a graduated tube or eudiometer. Unfortunately Frankland's only eudiometer had exploded a few weeks before and he was now about to go for a second time to Marburg. Packing his sealed tube securely in a trunk he made an uneventful voyage to Germany, arriving there in October with his sealed tube (incredibly) still intact. For various reasons the opening had to wait until the following February, but when it did take place the results were as dramatic as they were unexpected. An immediate and violent reaction took place. The liquid that was originally present in the tube disappeared entirely and in its place was formed a large volume of inflammable gas. Moreover, white crystals present in the tube reacted with water to give zinc iodide and a further gas.

Frankland used various standard techniques of gas analysis. He measured the changes in volume on combustion, comparing these with values expected for pure gases; he removed alkenes by absorption in sulphuric acid and noted the diminution in volume; he tested for homogeneity by comparing samples of gas before and after diffusion and, of course, he deduced gas densities.

It became apparent that large quantities of butane were truly formed and Frankland's dream of 'isolating ethyl' was fulfilled. Clearly a most important new reaction had been discovered and it has now become evident that the zinc had reacted initially with the ethyl iodide to form quite a new kind of compound: diethylzinc. It is this which reacts so violently with water to produce hydrocarbon gases. Further research made it clear that a new general reaction had been discovered between alkyl halides and zinc, and in the next few months Frankland produced the methyl and amyl analogues of diethylzinc. To this new type of compound he gave a new name: *organometallic*, which he defined as a compound containing a metal conjoined directly to carbon.

When Frankland returned to England in the winter of 1849–50 he determined to pursue further his investigations of organometallic compounds. Having obtained a teaching post at the then College of Civil Engineering in Putney he began work on various metals and alkyl halides, taking advantage of the space on the roof of his laboratory, unscreened by neighbouring buildings, where, as the spring advanced in 1850, he was able to expose his sealed tubes to the rays of the sun.

The consequences of this work for chemistry were momentous as they led directly to a theory of valency. But they also laid open a whole new area of chemical syntheses, for all of these organometallic compounds were highly reactive and thus able to play an important role in chemical synthesis. Ultimately those that Frankland discovered were replaced by the more amenable Grignard reagents, but that was many years later. Until that time they provided an important cluster of new general synthetic methods.

Figure 2.9 A sealed tube containing sample of Frankland's diethylzinc: note the large amount of unreacted metal.

In modern terms, the chief reactions effected by Frankland were these, C_2H_5ZnI being the white solid and $(C_2H_5)_2Zn$ the colourless liquid:

$$C_2H_5I + Zn = C_2H_5ZnI$$
$$2C_2H_5ZnI = (C_2H_5)_2Zn + ZnI_2$$
$$C_2H_5ZnI + C_2H_5I = C_4H_{10} + ZnI_2$$
$$C_2H_5ZnI + H_2O = C_2H_6 + Zn(OH)I$$
$$(C_2H_5)_2Zn + 2H_2O = 2C_2H_6 + Zn(OH)_2$$

Now it must be obvious that these new compounds occupied a very dubious position in the chemical hierarchy. In some respects they seemed to be genuinely organic in their behaviour, but in others the presence of the metal conferred a very inorganic quality upon them. All that we need to note here is that Frankland's organometallic compounds by their very existence enabled the sharp demarcation lines between the two chemistries to be blurred if not obliterated. Later in this Course *The Nature of Chemistry* (Units 22–24) you will have further occasion to study the chemistry of organometallics and you will see that they still occupy a very ambivalent position in the structure of our science.

SAQ 2 (*Objective 3*) In what ways could this research of Frankland have contributed to the 'downfall' of electrochemical dualism?

SAQ 3 (*Objective 3*) Compare the motivations, strategy and achievements of this synthetic work of Perkin and Frankland.

2.1.3 The meaning of 'synthesis'

Now it must be fairly obvious from the preceding account that the term 'synthesis' has been bandied about in a wide variety of ways and with several meanings. Before attempting to assess how far 'synthesis' did or did not contribute to the unification of chemistry in the 1860s, it will be as well to clarify the various shades of meaning in the term, both in 19th century and modern usage.

Apart from having a general sense of a 'building-up',* synthesis has been used in chemistry with a great number of different meanings. In the definitions below, the term 'natural product' means a compound identical in every way with one occurring in nature; in contrast, an 'artificial product' is to be taken as one that, so far as is known, is not naturally occurring but can be produced in the laboratory. Synthesis can then have any of the following meanings:

1 Formation of a natural organic product from any other simpler natural product.

2 Formation of a natural organic product from an artificial product.

3 Formation of a natural organic product from the elements directly.

4 Formation of an artificial organic product from any other artificial product.

5 Formation of an artificial organic product from the elements directly.

6 Formation of an artificial organic product from a natural one of simpler structure.

It would of course be possible to have several of these products in series, e.g. the sequence 5, 4, 2, 1.

> In which of these senses would a synthesis have been expected to play a part in overthrowing vitalism in organic chemistry?
>
> ---
>
> Definitions 2 or 3, because only these led to a natural product from the non-organic starting materials.

The following are examples of reactions for which, at one time or another, a 'synthetic' status has been accorded:

(a) Wöhler's 'synthesis' of urea (Unit 1, Section 1.2.1.3).

(b) Kolbe's sequence of reactions leading to acetic acid (Section 2.1.2).

(c) The production of urea by the action of phosgene on ammonia (John Davy, 1811):

$$COCl_2 + 2NH_3 \longrightarrow CO(NH_2)_2 + 2HCl$$

(d) Production of oxalic acid, $(COOH)_2.2H_2O$, by oxidation of sucrose, $C_{12}H_{22}O_{11}$ (Scheele, 1766).

(e) Frankland's production of butane from zinc and ethyl iodide (1849; see also Section 2.1.2):

$$Zn + 2C_2H_5I \longrightarrow C_4H_{10} + ZnI_2$$

(f) Formation of benzoyl cyanide from benzaldehyde ('oil of bitter almonds'; see Unit 1, Section 1.2.3.3).

*It was used in that general sense by John Dalton in 1808: 'Chemical analysis and synthesis go no farther than to the separation of particles from one another, and to their reunion'.

(g) Perkin's production of 'aniline purple' from aniline (Section 2.1.2).

(h) Berthelot's formation of acetylene from carbon and hydrogen (Section 2.1.1).

(i) Berthelot's formation of methane from carbon disulphide (Section 2.1.1).

(j) Berthelot's formation of benzene by polymerization of acetylene (Section 2.1.1).

(k) Berthelot's production of 'artificial glycerides', i.e. synthetic fats, obtained by allowing glycerol to esterify with long-chain carboxylic acids.

(l) Lemieux' and Huber's production of sucrose from derivatives of glucose and fructose (1953).

> SAQ 4 (*Objective 4*) With which of its meanings 1–6 can the term 'synthesis' be applied to each of these examples (a)–(l)? Can you identify any special difficulties in coming to a conclusion?

It should not be assumed that a definition of 'synthesis' is a purely academic one belonging to the 19th century. In 1958 a tribunal was set up to consider whether certain polymers were properly excluded from the Key Industry Duty List and therefore liable to import duty. The issue hung upon whether these could rightly be classed as 'synthetic organic chemicals' or not, for, if they could, they could be protected from such duty. In the end the tribunal concluded that the polymers were properly excluded and remained liable to an import duty of 10 per cent. But the point of interest to us is that during the enquiry the following remarkable opinion was reported from one very distinguished chemical witness: 'that the nature of chemical compounds had been rather clear since 1808, and then the meaning of synthesis, he thought, had been rather clear since about 1840 or 1850, and it could be put back 10 years behind that. 'Organic' got its modern meaning about the same time, about 1832; he thought the whole thing had hardly changed since 1840.' (*Chemical Age*, 29 November 1958, p. 902). Which is a salutary reminder not only that chemists can be blissfully unaware of the history of their subject but also of the complexity of such a term as 'synthesis'.

The most important distinction to bear in mind today is that between *total synthesis* (senses 2 or 3 above) and *partial synthesis* (sense 1). Otherwise 'synthesis' still means a quite general 'building-up' of a molecule.

This analysis of 'synthesis' may alert us against sweeping generalizations about the role of synthesis in chemical thinking, both past and present. It is a very broad term, and, as we have seen, only in senses 2 and 3 does it have an anti-vitalist importance. We shall call a work undertaken with this objective *ideological synthesis* in order to distinguish it from the pursuit of new reactions with a more severely practical goal which we can designate as *technical synthesis*. Although these terms are our own, the distinction between the two approaches was very apparent in the mid-nineteenth century. There is no doubt that the English chemist William Odling, addressing the British Association in 1862, spoke for many chemists when he said:

> Despite much that had been done by Wöhler, Kolbe, and others, the full recognition of synthetic organic chemistry which now obtains, together with the great advances therein that have recently been made, is mainly due to the labour of Berthelot, prosecuted uninterruptedly for the last ten years.

Berthelot himself agreed with this assessment of his rivals' influence, arguing that the products of their syntheses lacked the general impact of his own:

> These syntheses are extremely interesting; but, by reason of the very nature of the bodies upon which they bear, they have remained isolated and without fruitfulness.

To some extent this seems to have become almost a matter of national pride. Certainly the French chemists were vociferous in their claims that vitalism was vanquished and little remained to prevent the union between organic and inorganic chemistry. It is noteworthy that Odling, whose words we have noted above, was at that time very sympathetic to the French School. Similarly, in a paper of

Figure 2.10 William Odling (1829–1921), English chemist noted for contributions to the early debates about valency. Though recalled today mainly for his incorrect formula for bleaching powder, he played a useful part in chemical teaching at London and, from 1872, as Professor at Oxford.

1862, C. A. Wurtz discussed how ethylene might be 'the connecting link between organic and mineral chemistry'. (Some years later Wurtz displayed his nationalistic passion in this classic example of how *not* to begin a textbook of chemistry: 'Chemistry is a French science, it was constituted by Lavoisier of immortal memory'.) Later we shall find Dumas asserting similar claims for unification.

However, in Germany a great new industry was developing based upon organic synthesis of the other kind: the manufacture of dyes derived from coal-tar. This industry had its origins in the pioneering work in the late 1840s and early 1850s at the Royal College of Chemistry in London, where Hofmann and his students attacked with great determination the problems associated with the complex mixture of substances that we know as coal-tar. It would take us too far from our purposes to be diverted into an account of those researches, but we can remind ourselves that the first synthetic dyestuff was discovered in London by W. H. Perkin, acting on a hunch given to him by Hofmann. Thereafter the pace of discovery accelerated tremendously and all efforts were bent to achieve better, brighter, faster, more beautiful dyestuffs on one hand and to exploit the dazzling complexity of the chemistry of the new aromatic compounds that were appearing almost every day. It is hard to imagine a greater difference of outlook and practice between this synthetic work and the ideological syntheses of Berthelot. In fact, many of Berthelot's own syntheses suffered from two major disadvantages.

SAQ 5 (*Objective 2*) Can you identify (a) a theoretical and (b) a practical disadvantage in Berthelot's syntheses from the elements?

Thus it is not surprising that J. R. Partington could assert that 'modern organic chemistry stems from Frankland and Kolbe rather than Berthelot'. Some of the consequences of these developments will appear in Section 2.4 and in Unit 3. As we take our leave of Berthelot we must not minimize his great contribution to the chemistry of fats and much else besides. Nor is it true to say that his ideological syntheses were without significant effects on the unification of chemistry. They undoubtedly helped to reinforce the convictions that had originated in the analogy drawn by Berzelius. But so much was happening in the 1860s that it would be foolish to try to isolate—and therefore magnify—their long-term effects. To some of these other developments we must now turn.

Figure 2.11 C. A. Wurtz (1817–1884), French chemist who worked chiefly in Paris and is famous for his studies of ethylene glycol, ethylene oxide and ethylamine. He was a founder of the Société Chimique de France which became the centre of French organic chemistry.

Figure 2.12 August Kekulé (1829–96), German chemist who, with Frankland and others, may be said to have founded the theory of valency and, with Butlerov, to have created the theory of structure. The most spectacular achievement he accomplished was undoubtedly his famous structure of benzene. After an assistantship at St Bartholemew's Hospital, London (1854–55) he moved to Heidelberg, becoming eventually Professor at Ghent (1858) and Bonn (1867).

2.2   The Karlsruhe Conference

2.2.1   The central problem facing chemistry in 1860

From the ambivalent effects of organic syntheses on the structure of chemistry in the 1860s we must now consider one of the great turning-points of chemical history. This was a conference at Karlsruhe held in 1860, the same year in which Berthelot published his book on synthesis. Convened in response to a most urgent need for agreement among chemists it became almost the first international scientific conference; in that respect alone it was important for setting a fashion that has, happily, continued to this day. At the invitation of the German chemist Kekulé 140 people attended from Germany, France, Britain, Russia and elsewhere, including most of the leaders of chemical thought in Europe. The purpose of the meeting was to come to some common view on such basic concepts as atom, molecule, equivalent, etc., to agree on values of equivalents and to initiate a reform in chemical notation and nomenclature. By any standards this was an ambitious programme for a mere three-day conference, particularly as chemistry was deeply divided into numerous factions, often on national lines, and personal feelings ran high. So true was this that Kekulé did not dare agree to the nomination of a permanent chairman lest jealousies should be roused and the success of the conference be put at risk. As things turned out it nearly did end in failure; the way in which this disaster was averted will be seen later.

The holding of the Karlsruhe Conference is one of the clearest indicators imaginable of the state of despair into which chemistry had fallen. Despite all the success of the organic chemists the more they discovered the more baffling did the situation become. Penetration of Wöhler's primeval forest had indeed revealed a trackless jungle, and one of far greater extent than anyone had realized. A passage could be hacked through it only with the aid of a consistent and generally agreed atomic theory, and this they did not have. How could one possibly gain any real insight into the nature of, for example, acetic acid if one had no idea how many atoms of each kind were combined together within it?

Inorganic chemistry was, if anything, in an even worse plight. Progress here was slow and painful, partly, at least, because the central problem was more prominent. This was *the uncertainty surrounding all atomic weights*. An organic chemist

Figure 2.13 The Diet building at Karlsruhe, in the Grand-Duchy of Baden, where the Conference was held in 1860.

could at least make certain assumptions about the atomic weights of carbon, oxygen, etc., and develop some kind of a self-consistent scheme, but every time an inorganic chemist encountered another element he had to face the problem all over again. The only safe thing to do was to regard atomic weights as unknowable and to concentrate on what one *could* measure: combining or equivalent weights. Even that had its difficulties when an element had more than one equivalent (e.g. combined with oxygen in two or more different proportions by weight). Such uncertainties were reflected in formulae. Just before this time even a simple substance like water was represented in at least five different ways, as below. The top three are effectively the same (the barred symbol $\bar{H}$ being Berzelius' shorthand for two atoms of hydrogen), but the other two are quite different:

$$H_2O \qquad H^2O \qquad \overline{HO}$$

$$HO$$

$$H_2O_2$$

> Given that 8 g of oxygen combine with 1 g of hydrogen to form water, what atomic weights for oxygen $(H = 1)$ are implied by these formulae?

> For the first three $O = 16$; for the last two $O = 8$.

The last formula indicates a further complication. Even given atomic weights there was no notion at all as to what actual numbers of atoms might be involved in a molecule—only their ratios. On the basis $O = 8$ one might have had $H_{100}O_{100}$ just as well. One needed to know additionally the molecular weight and this was the crucial missing item; it was in fact more important than the atomic weights in some respects, as we shall see.

With a dilemma of this magnitude facing the conference it is less than surprising that it made heavy weather of its discussions. At the very first session of a steering committee a deep cleavage of opinion was revealed on the relevance of physical methods to chemical problems. Since this issue turned out to be the crux of the problem we must look at it a little more closely.

2.2.2 A physical answer to a chemical problem?

It will be recalled, perhaps, that in 1809 Gay-Lussac had shown that there was a simple whole number relationship between the volumes of combining gases and that of their product (if gaseous). Here are some well-known examples:

(a) 1 vol. of hydrogen+1 vol. of chlorine $\longrightarrow$ 2 vol. of hydrogen chloride

(b) 2 vol. of hydrogen+1 vol. of oxygen $\longrightarrow$ 2 vol. of steam

(c) 3 vol. of hydrogen+1 vol. of nitrogen $\longrightarrow$ 2 vol. of ammonia

(d) 1 vol. of carbon monoxide+1 vol. of chlorine $\longrightarrow$ 1 vol. of carbonyl chloride

(e) 2 vol. of ethane+7 vol. of oxygen $\longrightarrow$ 4 vol. of 'carbonic acid'+6 vol. of steam

At first sight there seems to be no regularity; sometimes there is no change of volume, sometimes a net expansion and at others a net contraction. Yet the underlying simplicity cannot be fortuitous; it must mean *something*. The problem was 'what?'.

The simplest interpretation is that equal volumes contain equal numbers of particles. This possibility had been entertained, but rejected, by Dalton. If one considered case (a) in that light one would have:

1 particle of hydrogen+1 particle of chlorine $\longrightarrow$ 2 particles of hydrogen chloride

So far so good. But if the question is now asked 'How many particles of hydrogen will give *one* of hydrogen chloride?' the uncomfortable answer is of course one half. Perhaps scared off by the prospect of dividing an atom ('Thou knowst thou canst not split an atom') the Quaker chemist would have nothing to do with Gay-Lussac's law; he simply disputed its accuracy!

A solution to the problem was proposed by the Italian physicist Amadeo Avogadro in 1811, distinguishing between '*molécules intégrantes*' (our molecules) and '*molécules élémentaires*' (our atoms). Equal volumes of all gases will then contain equal numbers of molecules, not atoms. If this is true the reply to the question above is 'half a molecule' which is quite tenable if that molecule contains an even number of atoms.

It is important to note why this is called Avogadro's 'hypothesis' rather than his 'law'. Unlike Gay-Lussac's law, which was a generalization based upon observation, Avogadro's proposals were strictly in the realm of theory; no one had any evidence at all that could point *unambiguously* to what Avogadro said. Many years later it became possible to deduce something like his hypothesis from another theoretical scheme, the kinetic theory of gases. Its probability is now very high indeed but even today it is preferable to recognize it as a 'hypothesis'.

Now let us see how Avogadro's hypothesis could be of value in determining molecular weights. To use it we require to measure the *density* of a gas X and compare it with that for hydrogen; let these be respectively D_X and D_{hyd}. Then

$$\frac{D_X}{D_{hyd}} = \frac{\text{mass of given volume of X}}{\text{mass of same volume of hydrogen}}$$

If the molecules of X and of hydrogen have molecular weights respectively M_X and M_{hyd} and, by Avogadro, there are the same numbers of molecules in both gases since their volumes are identical,

$$\frac{D_X}{D_{hyd}} = \frac{M_X}{M_{hyd}}$$

Since the densities can be readily measured one can determine the molecular weight of a gas in terms of that for hydrogen. Evidence that the hydrogen molecule contained an even number of atoms came from reactions like (a) above; since no evidence appeared to suggest that this number should be more than 2 the hydrogen molecule was accepted to be H_2 and its molecular weight was therefore 2. Hence the expression above becomes:

Relative density $(D_X/D_{hyd}) = \frac{1}{2}$ molecular weight

One way in which this relationship, based upon Avogadro's hypothesis, could be used in the determination of atomic weights was as follows. Relative densities would be measured for as many volatile compounds containing an element as possible, and these would be taken in conjunction with the percentage weight analyses of the element concerned. The atomic weight would be then taken as the smallest 'contribution' of that element to a molecular weight of its compound. Thus consider these data for some compounds of carbon:

Compound	Relative density of vapour	% of carbon
carbon disulphide	38	15.8
carbon dioxide	22	27.3
methanol	16	37.5
acetic acid	30	40.0
carbon monoxide	14	42.0
ethanol	23	52.2
methane	8	75.0
ethane	15	80.0
acetylene	13	92.3
benzene	39	92.3

SAQ 6 (*Objective 5*) Using this method calculate the probable atomic weight of carbon, and indicate the assumptions made.

Figure 2.14 Amadeo Avogadro (1776–1856), Italian physicist most noted for his famous 'hypothesis', propounded in 1811 but largely ignored until its revival by Cannizzaro nearly fifty years later. He wrote a well-known textbook of physics and introduced the metric system into Italy. His Chair of Physics at Turin was suppressed in 1823 following the political unrest raging through Italy at that time, but he was restored to it in 1834.

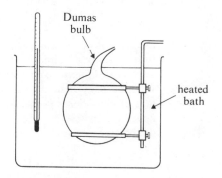

Figure 2.15 Dumas' vapour density apparatus: a thin glass bulb, containing the liquid to be vaporized, is immersed in a bath well above the boiling temperature of the liquid. Air is expelled through the narrow tube, which is then sealed off. From the weight of bulb the weight of vapour is easily obtained, while its volume is determined by subsequently opening the bulb and then filling with water and reweighing. Hence its density is found and, after suitable corrections for temperature and pressure, the vapour density can be calculated.

2.2.3 An alternative approach

This then is how Avogadro's hypothesis could have been, and eventually was, used. It was to be the crowning achievement of the Karlsruhe Conference that this method was restored to chemistry, for, in sad truth, it had virtually disappeared from view for nearly half a century. It is necessary to ask why.

On the basis of what you have read in this Course so far, can you suggest possible reasons?

We have already noticed the opposition of Dalton, and his authority was widely respected. More important was the rejection of the hypothesis by Berzelius, because if applied to products of reactions it implied double molecules like H_2 and O_2, which his electrochemical theory could not tolerate.

There was more than that to it, however, although we must not underestimate the importance of Berzelius' qualified rejection. By applying Avogadro's ideas to reacting gases only, Berzelius did at least arrive at consistent (and in modern views correct) atomic weights for oxygen, nitrogen and some other common elements. Some people have seen a reason for Avogadro's rejection to lie in his personal circumstances. He was little known in Italy (which was, in any case, being torn by revolution). His paper was marred by an unjustifiable extension of his views to non-gases. Yet when Ampère put forward a similar theory in 1814 it was similarly ignored for the most part, though he was both famous and French.

More important, perhaps, was the accumulation of vapour density data that were very hard to square with Avogadro's hypothesis. The pioneer in vapour density techniques in the first half of the century was Dumas himself, and some of his results, despite their accuracy, were quite inexplicable. Thus consider these figures for four elements sufficiently volatile at high temperatures for their vapour densities to be measured. Only in one case did their densities agree with what would have been expected on the basis of other atomic weights obtained by other methods (see below) and assuming all molecules were diatomic:

Element	mercury	iodine	phosphorus	sulphur
Found	100	127	68	94
Expected	200	127	31	32

Can you see why this disparity should exist?

In fact these elements all have different numbers of atoms in their molecules, which are Hg, I_2, P_4 and (for sulphur) an average approximating to S_6.

Dumas was not to know that and about this time (the early 1830s) the atomic theory itself was coming under fire for its inability to accommodate such awkward facts. Later workers noted other baffling results; for example, the observed vapour density of ammonium chloride was only half of what was expected, while that of phosphorus pentachloride decreased with rise in temperature. Both we now know to have been due to a dissociation but at the time they helped to cast doubt on any so-called 'physical' method for determining molecular weights:

$$NH_4Cl = NH_3 + HCl$$
$$PCl_5 = PCl_3 + Cl_2$$

It would take us much too far off course to pursue the question of alternative approaches to atomic weights before Karlsruhe. One example must suffice, the case of Berzelius himself. Having amassed as much analytical evidence as possible from a wide range of compounds, especially oxides, he worked within four main guidelines.

1 Berzelius' own 'corpuscular theory'

This was a tentative and empirical set of rules about how pairs of atoms are likely to combine. Possible combinations would be:

$$A+B, \quad A+2B, \quad A+3B, \quad A+4B, \quad etc.$$
$$2A+3B, \quad 2A+5B, \quad 2A+7B, \quad etc.$$

and the simplest would always be the most likely. Moreover, if two elements combine in more than one way certain predictions can be made; in oxides the number of oxygen atoms combining with the same amount of the other element X in each will be in simple ratios, as 1 : 2 or 1 : 1½. This would give formulae like XO and XO², or XO and X²O³, or (if lower oxides were also expected, as for sulphur and iron) XO² and XO³. When a basic and acidic oxide combine in a salt a simple whole number relationship must exist between the number of oxygen atoms in acid and base. These 'rules', combined with superb experimental technique, enabled him to conclude, for example, that 'chromic acid' was CrO³ and its green analogue Cr²O³.

2 The law of isomorphism

This discovery by Berzelius' student Mitscherlich (Unit 1, Figure 21) was used to supplement other results. Thus, as the green chromium oxide was isomorphous with ferric and aluminium oxides these were designated respectively Fe²O³ and Al²O³. Hence, since their composition by weight was easily determined, the atomic weights would readily follow.

3 The law of atomic heats

In 1819 Dulong and Petit had suggested that, for a solid element, the product of its atomic weight and specific heat* was roughly equal to 6. Since specific heats were readily determined the *approximate* atomic weight could be obtained and used to show which of several experimental values was correct. Here is an example. In its more usual (black) oxide, copper combines with oxygen in the ratio 63.5:16; at one time it had been commonly assumed that the formula was CuO², in which case the atomic weight of copper would be 127 (assuming O = 16). But its specific heat was approximately 0.088, so the atomic weight will be about 63, i.e. 63.5. This led Berzelius to the correct formulae for many oxides, but he came to the erroneous conclusion that *all* strong bases were similarly constituted as XO. Despite disagreement with the atomic heat law he included the oxides of sodium, potassium and silver, which should have been Na_2O, K_2O and Ag_2O. This last mistake was particularly serious because one would often determine the molecular weights of organic acids by making their silver salts and then burning a known weight of them and measuring the amount of silver left. Given an atomic weight of silver that was twice the correct value one inevitably inferred a molecular weight for the acids that was also twice too high—and that is why Berzelius' formulae for them are wrong in that respect. You will recall that it fell to Gerhardt to halve such formulae on chemical grounds, so giving them the values of today (Unit 1, Section 1.3.1).

4 Theory of gaseous volumes

As we have seen Berzelius used volumes of gaseous reactants to come to correct formulae for compounds like H_2O and NH_3, and was thus able to secure correct atomic weights, especially important for oxygen. Had he been able to take the final step advocated by Avogadro there is little doubt that he would have advanced the cause of chemistry by half a century.

SAQ 7 (*Objective 5*) The following data were known to Berzelius. Using his methods determine the atomic weights on the basis O = 16.00.

(a) *Manganese* Three oxides are known, with these compositions by weight (%):

	I	II	III
manganese	78.07	70.34	64.03
oxygen	21.93	29.66	35.87

Also, the oxide II is isomorphous with ferric oxide.

(b) *Nitrogen* 177.26 parts by weight of nitrogen combine with 100 parts of oxygen, but their combining ratios by volume are 2 : 1.

*In modern terms, specific heat = the thermal energy that must be transferred to a mass of material to raise its temperature by 1 K, divided by the thermal energy that must be transferred to the same mass of water to raise its temperature by 1 K.

2.2.4 The rediscovery at Karlsruhe

In these ways Berzelius was able to arrive at atomic weight values for about 30 elements. Posterity has judged him to be right—astonishingly so—for all except the alkali metals and silver. But his contemporaries were less enthusiastic than posterity and, for many years, chemistry was ravaged by doubts and plagued by all manner of alternative schemes from the credible to the ludicrous. One key

Figure 2.16 Interior of the Second Chamber in the Diet building at Karlsruhe.

Figure 2.17 Stanislao Cannizzaro (1826–1910), Italian chemist, teaching successively in the universities of Genoa, Palermo and Rome. He is chiefly remembered for the reaction named after him (in which benzaldehyde is converted by alkali into benzoic acid and benzyl alcohol) and his revival of the hypothesis of his fellow-countryman Avogadro.

was missing that would have made sense of the whole riddle, and that was to lie unrecognized in the paper by Avogadro, long buried in the pages of the 1811 volume of the *Journal de Physique*. It was not to be properly exhumed until the Conference at Karlsruhe, and then not until the very last moment. It happened like this.

Prominent among the participants at the Karlsruhe Conference was a 32-year old Italian chemist, Stanislao Cannizzaro. Early in the proceedings he found himself in opposition to Kekulé on the admissibility of physical evidence for chemical molecular weights. A lone convert to the ideas of Avogadro he had for the past few years been teaching these to his chemical classes at Genoa, showing that despite the difficulties here was a better solution to the problems vexing all chemistry than any so far advanced. At Karlsruhe came the first real opportunity to put his case in public debate. After several clashes earlier in the Conference his speech at the final session, delivered with clarity and vigour, could have become the moment of truth for all present. But it was not to be. Instead the discussion descended to the level of a procedural wrangle and the Conference terminated on a note of total indecision.

Yet all was not lost. As the members were taking a frosty leave of one another they were given a small pamphlet written by Cannizzaro two years previously: *Sunto di un Corso di Filosofia chimica* (*Sketch of a Course of Chemical Philosophy*) which was a reprint of a paper in the Italian chemical journal *Il Nuovo Cimento*.

22

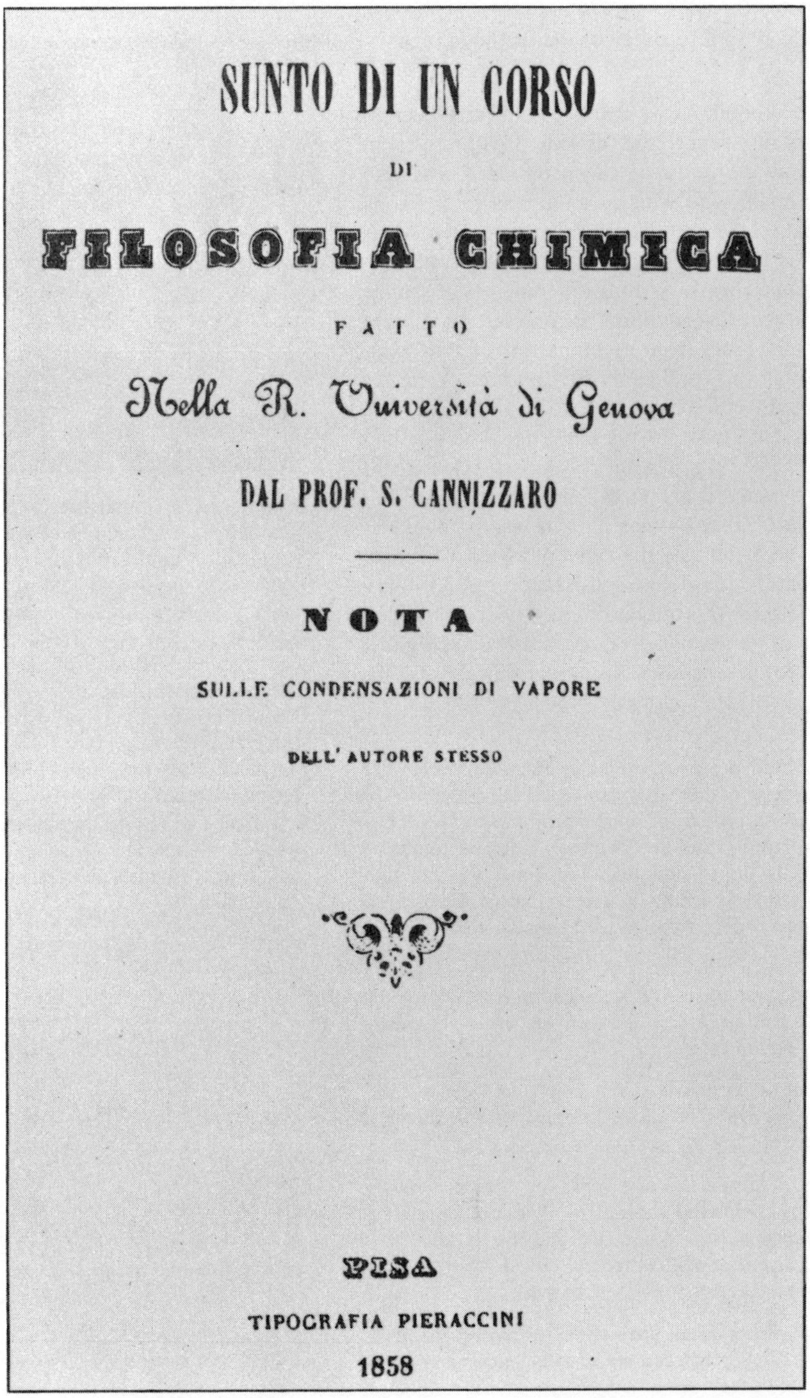

SUNTO DI UN CORSO

DI

FILOSOFIA CHIMICA

F A T T O

Nella R. Università di Genova

DAL PROF. S. CANNIZZARO

―――

N O T A

SULLE CONDENSAZIONI DI VAPORE

DELL' AUTORE STESSO

PISA

TIPOGRAFIA PIERACCINI

1858

Figure 2.18 Title page of Cannizzaro's *Sunto di un corso di filosofia chimica* (1858).

It was a summary of Cannizzaro's lecture course at Genoa and, with tables of results, it must have been a clearer exposition than his verbal presentation could ever have been. Unlike modern conferences, at Karlsruhe there were no prepared papers or visual presentation of data. Doubtless, as today, the free reprints were often disposed of unread, but in a few cases they made an impact that was to lead to the transformation of the whole science. One responsive recipient was the German research worker from Breslau, Lothar Meyer. On the way home he pulled the pamphlet from his pocket:

> I read it again and again and I was amazed at the clear light which that little paper shed on the main subjects of our debates. The scales fell from my eyes, doubts disappeared and a feeling of calm certainty took their place. If I was able to help in clearing up the points at issue and cooling the hot tempers, I owe much to Cannizzaro's pamphlet. Many other members of the Conference felt the same. The tides of battle began to ebb; the old atomic weights of Berzelius once more came into their own. After the apparent discrepancies between the laws of Avogadro and Dulong and Petit had been explained by Cannizzaro, both could be used to the full.

When, in 1864, Meyer's influential book *Die modernen Theorien der Chemie* was published chemistry was well on the way to a recovery of the methods of Avogadro and Cannizzaro.

The importance of this advance was profound. As we shall see in the next Section it led to a consolidation of the theory of valency, and, as Unit 3 will tell, to an enunciation of the Periodic Law. But the unity of chemistry itself was also fostered in a direct way. Of all the remarks to which Cannizzaro took strong exception perhaps the most provocative—as well as the most perceptive—was one from Dumas, in the chair for the last meeting. Speaking of the confusion then existing, with inorganic and organic chemistry using different sets of atomic weights, he argued that one should accept this situation because, after all, *there were two chemistries*. This statement in 1860, from a compatriot of Berthelot, should make one pause before asserting too confidently that by now chemistry was united; a deep division undoubtedly still existed. But for Cannizzaro such assertions were dangerous heresy, and he took up the challenge where it had been thrown down and vigorously affirmed that chemistry was one science, not two. For him the use of Avogadro's hypothesis had already bridged the gap, and it is interesting to note, nine years later, that Dumas himself was prepared to make the following admission. The occasion was the delivery of the Chemical Society's first Faraday lecture—a tribute to the great English natural philosopher who had died two years previously. Though this quotation may seem rather long it is worth reading carefully because of its chemical importance and the light it sheds on the attitude of many 19th century chemists. Speaking of organic compounds, he said:

> With regard to them, chemistry need no longer hesitate. Such substances are formed in the same manner as mineral matter, they exhibit all the conditions of its composition, its structure and its properties. One circumstance alone distinguishes them, and that, even, is not absolute. Generally, mineral matter is formed of simple elements, directly united, although sometimes we find that certain composite groups may act like elements, and may replace them or be replaced by them in combinations without the latter changing their general character.

> That which is the exception in mineral chemistry becomes the rule in organic chemistry. Lavoisier had the presentiment of this when he wrote—'Organic chemistry is the chemistry of compound radicals.'

> At this day, there is no longer any doubt. Cyanogen, cacodyl, and the metallic radicals of Frankland, are well known substances, offering all the chemical qualities of simple bodies, and nevertheless are compound.

> The law of substitution admits of displacing and replacing one element by another, or even, if desired, by a compound radical, in all the organic compounds of chemists, without changing their type. The converse is also true: there is nothing to prevent the replacement of a compound radical by an element in a combination; and this change again does not alter the type.

> Thus, the organic materials of which we speak are similar to mineral materials both in their nature and qualities: Firstly, they contain certain compounds which play the part of elements; Secondly, they are analogous to oxides, sulphides, chlorides; to acids, bases and mineral salts in all their properties; Thirdly, their radicals can replace the mineral elements, and be replaced by them; Fourthly, they can replace each other reciprocally.

> Like dead matter, these materials are susceptible of crystallisation, and of volatilisation, without becoming decomposed; they form definite combinations, are incapable of life, and never have lived. They thus resemble mineral matter in every respect, and they differ from it only by having as their constituents compound radicals, whilst the mineral species are generally compounds of elements.

> Analogies are apparent even in some of the numerical relations of the elements of mineral chemistry, and of the radicals of organic chemistry. These relations are manifest in both cases, for all elements which are capable of being arranged in series or natural families.

> Thus, lithium, sodium and potassium, whose respective atomic weights are 7, 23 and 39, form a series of which the difference is 16. Magnesium, calcium, iron, whose respective atomic weights are 12, 20, 28, form a series whose difference is 8.

> In the same way all organic radicals present the same relations. Methylium, ethylium, propylium, butylium, &c., give a series whose difference is 14, their respective atomic weights being 15, 29, 43, 57.*

Figure 2.19 J. Lothar Meyer (1830–95), German chemist and one of the first to respond to Cannizzaro's call for a return to Avogadro. He did so publicly and effectively in his *Die modernen Theorien der Chemie* of 1864. This contained the first hints of the periodic law which he and Mendeléev were later, and independently, to propose. At numerous university posts in Germany his research ranged widely over organic, inorganic and physical chemistry. He was also an able and articulate advocate of chemical education by research.

*i.e. methyl, ethyl, propyl and butyl.

24

It seems natural to conclude from all this evidence, not that substances thus endowed and composed are organic because they are derived from organised beings, but rather that the elements of mineral chemistry may be complex. In any case, the assimilation of the two chemistries becomes more and more close. . .

The function of the chemist changes, and becomes elevated. To his acquaintance with the numerous substances which his power evokes from the regions of the unknown, he owes his first attention to order, method, classification and nomenclature. But, once this primary duty is fulfilled, he contemplates the innumerable host of forms, raised by his conceptions, or realised by his hands; and he applies to mathematics for a definition of the harmonies of numbers therein revealed; to mechanics, for a statement of the laws which their structures obey, or those which determine the stability of the material systems which they represent.

The Bible tells us that when God had formed, from dust, all the beasts of the earth, and all the fowls of the air, he made them pass before Adam, and that the name given to each by Adam is its true name.

In presence of this new creation, not of animated beings—whose appearance on earth depends on a power superior to man's—but of the harmonious forms which chemistry reproduces at will, always like each other, and always distinct from all others, man might sometimes forget that he has to give names to the works of God, and only remember that he is naming the works of his own hands.

If the discoveries which we have witnessed during the last half century do not justify pride, they at least excuse it. But, to bring back man to the appreciation of truth, it suffices to tell him that—if he has become more expert in the art of observing, if he employs with more certainty the art of experimenting, if the logic proper to the sciences leads him more surely to the discovery of the laws of nature—he has not as yet advanced one step towards the knowledge of causes.

Let us consider, in particular, what he knows on the subject of the materials which his life sets in motion in its development, and the contrast will be striking.

If I question the physiologist, on the subject of these millions or milliards, of compounds, misnamed organic, which the chemist transforms, reproduces, or creates at pleasure, he will reply to the three following questions:—Are these compounds living?—No! Have they lived?—No! Are they capable of living?—No!

If I ask the chemist himself if these compounds belong to mineral chemistry—to the chemistry of dead matter—he will reply, Yes!

Organised matter, not capable of being crystallised, but destructible by heat, the only matter which lives, or has ever lived—this matter, a subordinate agent of the vegetating power in plants, of the motion and sensation of animals, cannot be produced by chemistry; heat does not give birth to it; light continues to engender it under the influence of living bodies.

Let us not be disturbed by a quibble. The ancients admitted that Nature alone produces organic matter, and that the art of the chemist is limited to transforming it. To-day we might, perhaps, pretend that chemistry is powerful enough to replace, in all respects, the forces of life, and to imitate its processes—let us, however, keep to the truth.

The ancients were mistaken when they confounded, under the name of *organic matter*, sugar and alcohol, which have never lived, with the living tissues of plants or the flesh of animals. Sugar and alcohol have no more share in life than bone-earth, or the salts contained in the various juices. These remnants, or rubbish of life, placed amidst organic matter, are true mineral species, which must be brought back to, and retained amongst, dead matter. Chemistry may produce them in the same sense that she manufactures sulphuric acid or soda, without, for all that, having penetrated into the sanctuary of life.

This subject remains what it was—inaccessible, closed. Life is still the continuation of life; its origin is hidden from us as well as its end. We have never witnessed the beginning of life; we have never seen how it terminates.

The existing chemistry is, then, all powerful in the circle of mineral nature, even when its processes are carried on in the heart of the tissues of plants or of animals, and at their expense; but it has advanced no further than the chemistry of the ancients, in the knowledge of life, and in the exact study of living matter; like that, it is ignorant of their mode of generation.

Where, then, is truly organised matter, or matter susceptible of organisation? What is its chemical constitution? What is its mode of production? What is its manner of growth?

Instead of myriads of species, one would feel disposed to recognise but eight or ten at most, if one may be allowed to consider elementary types of organisation as chemical species. Be this as it may, in the origin of beings which have life, we see cells appear, and in the heart of their types we find cells in place of organic elements, and still, beyond these, germs of cells.

In these cells, or in the spaces between them, we observe inert products, aliments, excretions, substances stored up. It is the cell, it is the germ which proceeds from life, which lives, which engenders life, and then dies. The substances which are contained in, or which surround these organs, are subordinate accidents, products rejected by organisation, or destined for its use, but distinct from life.

Every organised being is born of a germ; every plant from a seed; every animal from an egg. The physiologist has never seen the birth of a cell, excepting by the intervention or as the produce of a mother cell.

The chemist has never manufactured anything which, approximately or remotely, was susceptible of even the appearance of life. Everything he has made in his laboratory belongs to dead matter; as soon as he approaches life and organisation he is powerless.

Thus, for a century past, the empirical elements of matter have been recognised and separated; their combinations have been multiplied to infinity; physical forces have been brought back to a common origin—motion—and one has been at pleasure changed into the other; and yet—

Is the intimate nature of matter known to us? No! Do we know the nature of the force which regulates the movement of the heavenly bodies and that of atoms? No! Do we know the nature of the principle of life? No!

Of what use, then, is science? What is the difference between the philosopher and the ignorant man?

In such questions the ignorant would fain believe they know everything—the philosopher is aware that he knows nothing. The ignorant do not hesitate to deny everything; the philosopher has the right and the courage to believe everything. He can point with his finger to the abyss which separates him from these great mysteries,—universal attraction which controls dead matter, life which is the source of organisation and of thought. He is conscious that knowledge of this kind is yet remote from him, that it advances far beyond him and above him.

No! life neither begins nor ends on the earth; and if we were not convinced that Faraday does not rest wholly under a cold stone, if we did not believe that his intelligence is present here among us and sympathises with us, and that his pure spirit contemplates us, we should not have assembled on this spot, you to honour his memory, I to pay him once more a sincere tribute of affection, of admiration, of respect!

SAQ 8 (*Objective 13*) (a) How had Dumas' position changed?

(b) What basic distinctions was he now advocating?

(c) What conclusion does he draw about the elements?

SAQ 9 (*Objectives 5 and 6*) Figure 2.20 is a plate from Dalton's *New System of Chemical Philosophy* (1808); the numbers represent the 'simple' substances which are listed in the Key.

After inspecting these formulae please attempt the following questions:

(a) Can you identify any that are, by today's views, correct?

(b) What advances on this kind of formula had been made before Karlsruhe?

(c) What further advances were made just after Karlsruhe?

(d) What major change was still needed after that?

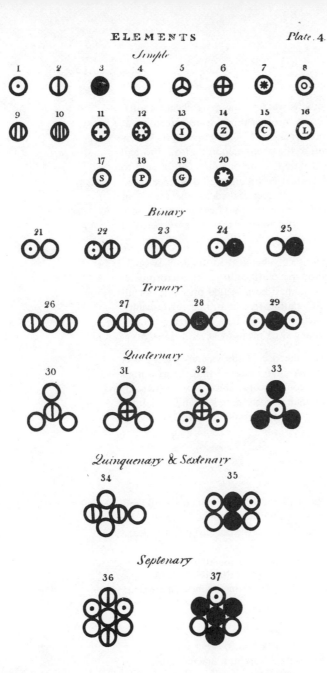

KEY

1 hydrogen
2 azote (nitrogen)
3 carbon
4 oxygen
5 phosphorus
6 sulphur
7 magnesia
8 lime
9 soda
10 potash
11 strontites
12 barytes
13 iron
14 zinc
15 copper
16 lead
17 silver
18 platina
19 gold
20 mercury

Figure 2.20 Plate from Dalton's *New System of Chemical Philosophy*, 1808.

2.3 The arrival of valency

As we have seen, by the early 1860s synthesis was making a significant contribution to the unification of chemistry. However, other things were happening at the same time which powerfully reinforced this effect. Not least among these new developments was the gradual recognition of the phenomenon that we now call valency. This is a complex idea and it is certain that no one man can be said to have discovered it in all its richness. But when eventually it did emerge it led directly to one of the most fundamental of all chemical ideas, the theory of structure. Even before this happened, however, the theory of valency by its very emergence from embryonic notions helped to cement a closer union between organic and inorganic chemistry.

Since valency is a concept that makes sense of chemical formulae it is fairly obvious that its usefulness was dependent upon a clarification of those formulae. In other words, until the atomic weight issue had been settled once and for all there would be little use in any doctrine relating to the atomic relationships within those formulae. Indeed, in this Section and the next we shall see how, within a few years, a chemical formula changed its meaning several times over.

Among those most closely associated with the rise of the valency theory was August Kekulé, and his own claims for being the founder of that theory sometimes led to an unfair devaluation of other people's work. Since, as we have said, the term 'valency' has, like synthesis, more than one meaning it would be helpful to look separately at the emergence of the separate strands of valency theory.

2.3.1 Saturation capacity

We have already paid some attention to the synthetic achievements of Frankland in his discovery and further study of organometallic compounds. It would take us too long to go into the details of his argument but his conclusions are clear:

> I had not proceeded far in the investigation of these compounds at Putney before the facts brought to light began to impress upon me the existence of a fixity in the maximum combining value or capacity of saturation in the metallic elements which had not before been suspected.

What Frankland had done was to observe that there was a limit to the extent to which an atom of, say, zinc or arsenic, could be attached to other atoms. As he himself said later: 'It was evident that the atoms of zinc, tin, arsenic, antimony, etc., had only room, so to speak, for the attachment of a definite and fixed number of atoms of other elements'. Accordingly, in the early 1850s, he proposed the following to the Royal Society (1852):

> When the formulae of inorganic chemical compounds are considered, even a superficial observer is struck with the general symmetry of their construction; the compounds of nitrogen, phosphorus, antimony and arsenic especially exhibit the tendency of the elements to form compounds containing 3 or 5 equivalents of other elements, and it is in these proportions that their affinities are best satisfied; thus in the ternal group we have NO_3, NH_3, NI_3, NS_3, PO_3, PH_3, PCl_3, SbO_3, SbH_3, $SbCl_3$, AsO_3, AsH_3, $AsCl_3$, etc.; and in the 5-atom group NO_5, NH_4O, NH_4I, PO_5, PH_4I, etc. Without offering any hypothesis regarding the cause of this symmetrical grouping of atoms, it is sufficiently evident, from the examples just given, that such a tendency or law prevails, and that, no matter what the character of the uniting atoms may be, the combining power of the attracting element, if I may be allowed the term, is always satisfied by the same number of these atoms.

Several things will be noted from this quotation. First, Frankland actually quotes from inorganic chemical compounds although his research was on the basis of organometallic substances. Secondly, he offers no hypothesis but merely draws attention to 'this symmetrical grouping of atoms'. Thirdly, he gives no list of saturation capacities. When he did give a few formulae they were, by modern standards, often wrong because his atomic weights were in error (in particular, $O = 8$).

Kekulé denied Frankland's claims on the grounds that his atomic weights were wrong. He also could not understand him because he (Kekulé) was an adherent of the type theory while Frankland remained in the radical tradition. Kekulé stood in the French tradition of Laurent and Gerhardt which deplored the dissection of molecules into radicals; Frankland owed more to Berzelius and did *not* regard his electrochemical dualism as totally discredited. But despite this ideological division, and from this distance of time, it certainly appears that Frankland must be credited with recognizing, however vaguely, the doctrine of saturation capacity.

2.3.2 Coupling together of polyvalent atoms

You will recall that adherents of the theory of types tended to write their formulae with curly brackets (see Unit 1, Section 1.3.2). But their purpose was limited. Said Gerhardt in 1856:

> Chemical formulae are not intended to represent the arrangement of atoms, but they are designed to make clear in the simplest and most exact manner relations which bodies manifest towards each other with respect to transformations.

They were artificial devices to indicate how the compound might react. In no sense were they meant to imply how the atoms really were arranged. In the 1850s a large amount of research in organic chemistry was leading members of this school to come gradually to the conclusion that the curly brackets might enshrine a truth deeper than they had once suspected. A classic case of this was that of A. W. Williamson at University College, London. His researches on the ethers led him to relate ether and alcohol to the water type in which one or more atoms of hydrogen were replaced by organic groups. It is hard to put a date to his enlightenment but there is no doubt that by the mid-1850s he was viewing these formulae in a new way. He was able to say that the oxygen atom *held together* the rest of the molecule. When he applied these ideas to more complex substances, e.g. sulphuric acid, it became evident that other groups than atoms could hold molecules together. He was careful not to be too definite in these ideas but they are clearly present in his writing. Here is an example. Sulphuric acid can be converted step-wise into chlorosulphonic acid and sulphuryl chloride.

Figure 2.21 A. W. Williamson (1824–1904), English chemist, Professor at University College, London, from 1849 to 1887. Another member of the group of London chemists in the mid-1850s whose discussions did much to clarify the emerging doctrine of valency, Williamson is now famous chiefly for his work on ethers. By means of the synthesis named after him ethers are made from alkyl halides and sodium derivatives of alcohols. This enabled him to perceive the relation between alcohols and ethers and to regard both as alkyl-substituted water molecules, i.e. as belonging to the 'water type'; he also saw that the oxygen atom holds together the rest of the molecule.

$$
\begin{matrix}
& H & & & H & & & \\
& C & & & O & & & Cl \\
SO_2 & & \longrightarrow & SO_2 & & \longrightarrow & SO_2 & \\
& O & & & Cl & & & Cl \\
& H & & & & & &
\end{matrix}
$$

> The existence of this body [chlorosulphonic acid] … furnishes the most direct evidence of the truth of the notion, that the bibasic character of sulphuric acid is owing to the fact of one atom of this radical SO_2 replacing or (to use the customary expression) being equivalent to two atoms of hydrogen.

You will notice that the curly brackets have disappeared, but the time had not yet come when they would be replaced by straight lines.

Other workers followed up Williamson's ideas, including Odling and Kekulé. It was the latter who most clearly pointed to the belief that atoms or groups of atoms could hold together parts of a molecule, and this particularly in his paper of 1857. But that was only the beginning. The most pressing problems of constitution lay in the field of organic chemistry and the next strand in the complex concept of valency is that of the tetravalency of carbon itself.

2.3.3 Tetravalency of carbon

During the 1850s it became urgently necessary to find a coherent theory to accommodate the rapidly growing numbers of organic compounds. In 1858 Kekulé published a paper entitled 'Ueber die Constitution und die Metamorphosen der chemischen Verbindungen und über die chemische Natur des Kohlenstoffs' (On the constitution and metamorphoses of chemical compounds and on the chemical nature of carbon). This celebrated document gave to the world the first clear assertion that carbon has a valency of 4. In tackling matters of great complexity in organic chemistry Kekulé said 'I deem it necessary ... to go back to the elements themselves that compose the compound'. Accordingly he considered typical formulae and emphasized in them the role of individual atoms of the elements. Thus consider his formula for benzenesulphonic acid:

$$\left.\begin{array}{c} C_6H_5 \\ SO_2'' \\ H \end{array}\right\}O$$

and compare it with the modern formula:

$$C_6H_5{-}SO_2{-}O{-}H$$

He wrote of 'the radical of sulphuric acid (SO_2'')' appearing to be 'diatomic' or, in our language, divalent, and holding together the rest of the molecule. He then came to the decisive pronouncement:

> If only the simplest compounds of carbon are considered ... it is striking that the amount of carbon which the chemist has known as the least possible, the atom, always combines with four atoms of a monatomic, or two atoms of a diatomic element; that generally, the sum of the chemical units of the elements which are bound to one atom of carbon is equal to four. This leads to the view that carbon is tetratomic [tetravalent].

Much has been written of Kekulé's immensely important contribution but it must not be allowed to overshadow the somewhat smaller one of the Scotsman Archibald Scott Couper. In 1858 he presented a paper 'On a New Chemical Theory' which was published in several journals and which had very clear formulae implying a tetravalent carbon atom. Unfortunately his thesis was weakened to some extent by a curious formulation for oxygen (writing it as two atoms) and his reputation was never established because of a total breakdown in health shortly afterwards. However, between them, Kekulé and Couper may undoubtedly be said to have given the world the concept of the tetravalent carbon atom.

2.3.4 Catenation of carbon atoms

For an organic chemist today there is one major element missing and this was provided by both Couper and Kekulé. It was the doctrine of catenation, i.e. the belief that carbon atoms can and do join themselves together in chains. These ideas were clearly expressed by Kekulé who calculated the number of 'affinity units' (valencies) available in a long carbon chain of n atoms to be $(2n+2)$. Couper actually gave formulae showing this kind of thing, and the following is an example of his representations of ethylene glycol:

Couper's formulae   Modern formula

His curious use of a 'double oxygen atom' should not be allowed to obscure his real achievement.

In this way the final thread in a concept of valency was put into place and it now became possible to apply this consideration to the whole range of chemical compounds.

In this highly compressed version of a most complex story we have merely indicated the four main elements. What perhaps is most remarkable of all is that the elegant and simple concept of valency had to emerge in such a devious and convoluted way. Surely, one may ask, it would have been more natural for the concept of divalent oxygen and monovalent hydrogen to emerge from the consideration of a simple molecule like water H_2O once that formula was

Figure 2.22 Archibald Scott Couper (1831–92), Scottish chemist who studied at Glasgow, Edinburgh and Paris, most famous for his 1858 paper 'On a new chemical theory', proposing tetravalent carbon and catenation. His promising career was prematurely ended shortly afterwards by the onset of severe mental illness.

30

established.* To think like that is to miss a very important point about the development of chemistry, which is that chemists tend to produce theories only in response to urgent pressures. Valency arose in response to the mounting pressures exerted by piles of accumulated, unsorted and unacceptable facts. After all, in the case of H_2O, what more was there to say than there were two atoms of hydrogen and one of oxygen? It was not necessary to import vivid mental pictures of atoms like billiard balls joined by springs or rods. Only as the science of *organic* chemistry expanded did the problems become acute. So it was that the vision of Dumas (Unit 1, Section 1.3.2) became true: organic chemistry really did 'give rules' to the rest of chemistry. Valency, universal in its application, emerged out of a study of the chemistry of one element, carbon. Even if we make an exception in Frankland's case, his work was still engaged within a framework of organic chemistry. Clearly valency could not become a definitive part of chemistry until atomic weights were universally agreed. Once that happened its unifying effect on chemistry could scarcely be overstated.

> SAQ 10 (*Objectives 7 and 8*) In 1877 Frankland wrote of the theory of saturation capacity announced in his paper of 1852: 'This hypothesis...constitutes the basis of what has been called the doctrine of atomicity or equivalence of the elements; and it was, so far as I am aware, the first announcement of that doctrine'.
>
> Four years later Kekulé wrote in a manuscript that was never to be published in his lifetime: 'I extended the idea of valency or atomicity of the radicals formulated by Williamson to the atoms of the elements also'. This is essentially a repetition of a claim made in 1864. 'Unless I am mistaken I am the one who introduced the idea of the atomicity of the elements into chemistry'.
>
> Can you offer some possible explanation for the great discrepancy between these two views? Neglect issues of personality and concentrate on matters of chemistry.

As we have now seen there were major items of principle which separated Frankland and Kekulé and made each incapable of recognizing fully the merits of the other's work. Pursuing the atomic weight issue for a moment, let us hear further the argument put forward by Kekulé:

> There is then the further consideration that most chemists in 1853 did not yet clearly distinguish the ideas of atom and equivalent, and that Frankland also did not make such a distinction, but spoke now of atoms, now of equivalents, whereas valency theory assumed a sharp difference between the two ideas.
>
> Frankland argued with a false atomic weight for oxygen, and had to be able to argue also for a false atomic weight for antimony, *etc.* His law does not even deal with atoms. As far as bodies are concerned, it denotes that they shall have a definite saturation capacity and deals with that relative quantity we regard as a radical.... The law refers not to atoms but equivalents. It is right for equivalents but false for atoms, and therefore false in the form arrived at by Frankland.

So for Kekulé Frankland has forfeited a right to be considered a founder of valency, not because he had the wrong atomic weights, but because he had *none at all*. He was not really talking about atoms, thought Kekulé, but only about equivalents (combining quantities). Whether one agrees or not it remains a powerful illustration of the relevance of the atomic weight issue to valency. Frankland's atomic weights were amended to modern values after Karlsruhe.

Even more fundamental than this, however, was the ideological division between those who held to the older radical views and those who subscribed to types. If this seems an exaggeration here is an assessment of the situation from someone who knew both men well—H. E. Armstrong:

> The problem first solved by Frankland was in the air—chemists everywhere had it in mind, especially in France. Kekulé was in London in 1854 and consorted with Williamson, who like himself was under the enthralling influence of Gerhardt. Had he consorted with Frankland, a man infinitely in advance of himself and most other chemists as a worker, his attitude would not have been so independent. I have not been able to discover that Kekulé made the least attempt to exchange views with Frankland, having joined another camp. We are in face of a psychological puzzle.

*Thus Williamson did not actually say that oxygen binds together the two hydrogen atoms in water until 1869.

The solution to that puzzle doubtless has something to do with the personal background and circumstances of the two men concerned. But it also has much to do with the differences of approach between those in the radical and type traditions. This is how Kekulé looked back on it in 1890:

> Fifty years ago, the stream of chemical progress had divided into two branches.... At length, as the two branches had again approached much nearer to one another, they were separated by a thick growth of misunderstandings, so that those who were sailing along on the one side neither saw those on the other, nor understood their speech. Suddenly a loud shout of triumph resounded from the host of the adherents of the type theory. The others had arrived, Frankland at their head. Both sides saw that they had been striving towards the same goal, although by different routes. They exchanged experiences; each side profited by the conquests of the other; and with united forces they sailed onward on the reunited stream. One or two held themselves apart and sulked; they thought that they alone held the true course,—the right fair-way,—but they followed the stream.
>
> Our present opinions do not, as has frequently been asserted, stand on the ruins of earlier theories. None of the earlier theories has been recognized by later generations as entirely false; all when stripped of certain ill-proportioned, meaningless excrescences, could be utilized in the later structure, and form with it one harmonious whole.
>
> Here and there a seed may have lain in the ground without germinating; but everything that grew came from seed that had been previously sown. My views also have grown out of those of my predecessors and are based on them. There is no such thing as absolute novelty in the matter.

This famous quotation is a standing invitation to all of us who study modern chemistry to beware of chemical exclusiveness. It is quite possible to become so absorbed in (for instance) the use of M.O. techniques that alternative approaches utilizing resonance are treated with some contempt, and vice versa. If the history of chemistry can teach us anything it is surely this point made by Kekulé, that truth is often found in a fusion of several different viewpoints and is rarely confined to one. This is why in this Course you will find a variety of approaches, each of which seems the best for the matter in hand.

So the theory of valency emerged from a fusion of the doctrine of radicals with the theory of types, and it was of course greater than either. This convergence of the two major schools of thought within organic chemistry was an important element in the even larger movement towards the unification of chemistry itself.

2.4 Structure

There are some ideas in chemistry that are uniquely characteristic of the subject, and others (like atoms and molecules) that it shares with other sciences. An example of the first category is the concept of *chemical individuality* which is closely related to the notion of *chemical purity*. Thus we assume that every pure sample of, say, ethylene will display identical properties with those of every other sample of the same chemical individual. In the 1860s another concept of this kind emerged, one that was in fact dependent on the notion of a definite chemical individual. This was the idea of *chemical structure*. So deeply has this become ingrained in our consciousness that chemists use it constantly without realizing that they are doing so. The theory of structure states that the properties of a compound depend entirely on the definite relationships existing between all the atoms of the molecule, the complete description of which is termed its 'structure'. It implies that that structure is uniquely important—much more so than, for instance, the origins of the samples, whether living or non-living. Moreover, it suggests that, for each compound, a unique structural formula can, at least in principle, be written.

It is obvious that the first realization of these matters must have made a profound impact upon chemical thinking generally. Coming as it did in the 1860s it adds further to the impression that this must have been an exciting time for chemists to be alive. And we may add that the chemistry of structure had a great effect on the structure of chemistry! But first we must notice the factors inhibiting such growth.

2.4.1 Opposition to structural ideas

It may seem surprising today that all chemists did not immediately welcome the theory of structure as one of the most important advances of their time. To us it explains so much and is such a powerful intellectual tool. Yet it took a considerable time to take root and opposition was at first intense.

> Can you suggest why this should be?
> _____
>
> We have already noticed the view of Gerhardt and his followers that formulae represent the way a molecule reacts *and nothing more*. This sprang from a perfectly correct reluctance not to go beyond the evidence and was particularly strong in France. Originated by Gerhardt in about 1839 it was still the view of Berthelot in 1877 (though by then he was almost on his own).

This characteristic reluctance of French chemists to think structurally is related to their general attitude to *atomic models*. It has often been pointed out that in France a mechanical model of a molecule seemed impossibly naïve, whereas few Britons or Germans would find it so. There may have been more to it than that. Quite often, it seems, an attitude of religious scepticism went hand-in-hand with a 'structural agnosticism'—and this was by no means limited to France, though a generally materialistic and positivistic philosophy of life was fairly characteristic of French thought in the mid-19th century.

There was another factor that restrained the rapid growth of structural ideas in the 1850s and that was well summed up by Kekulé in 1861:

> Many chemists, strange to say, are yet of the opinion that from the study of chemical change the constitution of compounds can be deduced with certainty, and therefore the position of the atoms can be expressed in chemical formulae. That the latter is not possible requires no special proof; it is self-evident that one cannot represent the positions of atoms in space—even if one had discovered them —by placing letters side by side in the plane of the paper, and that at least one requires a perspective drawing or model.

> It has sometimes been said that Kekulé did not really approve of the structure theory. Do you agree in the light of that quotation, and can you identify the point of difficulty?
> _____

33

Kekulé did not oppose the idea of structure, but he was afraid of unwarranted inferences about three-dimensional structure from formulae in one plane.

Here then was a danger of which chemists must always be aware—the natural tendency to go beyond the strict evidence and make unjustified extrapolations from the facts available. The fear that formulae might be taken to mean more than they were intended to mean was a powerful disincentive against using structural formulae too liberally. The extension of structure to three dimensions did take place in 1874, but the first structural formulae had no steric implications at all; the actual positions of atoms in space were unknown and unknowable.

In organic chemistry a further difficulty arose in the phenomenon later known as tautomerism.

SAQ 11 (*Objectives 9 and 12*) Why should this phenomenon be a difficulty in the growth of structure theory?

Some idea of the confusion attending the exercise of these doubts about valency and structure can be gained by looking at Figure 2.23. The list comes from Kekulé's *Lehrbuch*, published in 1861, and represents 19 different formulae for one compound, acetic acid, commonly in use in the late 1850s. It has sometimes been said that this represents one of the urgent reasons for the Karlsruhe Conference,

$C_4H_4O_4$ . . . . empirische Formel.

$C_4H_3O_3 + HO$ . . . . . . dualistische Formel.

$C_4H_3O_4$. H . . . . Wasserstoffsäure-Theorie.

$C_4H_4 \quad + O_4$ . . . . . . Kerntheorie.

$C_4H_3O_2 + HO_2$ . . . . . Longchamp's Ansicht.

$C_4H \quad + H_3O_4$ . . . . . Graham's Ansicht.

$C_4H_3O_2.O + HO$ . . . . . Radicaltheorie.

$C_4H_3 . O_3 + HO$ . . . . . Radicaltheorie.

$\left. \begin{matrix} C_4H_3O_2 \\ H \end{matrix} \right\} O_2$ . . . . . . Gerhardt. Typentheorie.

$\left. \begin{matrix} C_4H_3 \\ H \end{matrix} \right\} O_4$ . . . . Typentheorie (Schischkoff etc.)

$C_2O_3 + C_2H_3 + HO$ . . . Berzelius' Paarlingstheorie.

$HO.(C_2H_3)C_2,O_3$ . . . . . Kolbe's Ansicht.

$HO.(C_2H_3)C_2,O.O_2$ . . . . ditto

$\left. \begin{matrix} C_2(C_2H_3)O_2 \\ H \end{matrix} \right\} O_2$ . . . . Wurtz

$\left. \begin{matrix} C_2H_3(C_2O_2) \\ H \end{matrix} \right\} O_2$ . . . . . Mendius.

$\left. \begin{matrix} C_2H_2 . HO \\ HO \end{matrix} \right\} C_2O_2$ . . . . . . Geuther.

$C_2 \left\{ \begin{matrix} C_2H_3 \\ O \\ O \end{matrix} \right\} O + HO$ . . . . Rochleder.

$\left(C_2 \dfrac{H_3}{CO} + CO_2 \right) + HO$. Persoz.

$C_2 \left\{ \begin{matrix} C_2 \\ H \end{matrix} \right. \left\{ \begin{matrix} O_2 \\ H^2 \end{matrix} \right.$

$\dfrac{H}{H} \left\{ O_2^- \right.$ . . . . . Buff.

Figure 2.23 Page from Kekulé's *Lehrbuch der organischen Chemie* (1861) showing 19 different formulae for acetic acid.

but inspection of the table shows that *all* formulae are constructed on the same atomic weight basis, i.e. $C = 6$ and $O = 8$. The real problem was not seen as how many atoms were present but rather the nature of their arrangement. Each of these formulae represents insights gained from different reactions. They could therefore be said to conform to Kekulé's views of *rational formulae* (as opposed to those indicating permanent structure):

> Thus it will perhaps be possible to establish the constitutional formulae for compounds, which must naturally be unchangeable for one and the same substance. However, even if this is successful, different rational formulae (decomposition formulae) are still always permissible, because a molecule that is produced by atoms positioned in definite ways can be split in different ways under different conditions, and therefore can yield fragments of different sizes and compositions.

2.4.2 Emergence of the concept of structure

Although both Kekulé and Couper have been seen as authors of the theory of structure, and undoubtedly the idea was implicit in their writings, the actual phrase was introduced by the Russian chemist A. M. Butlerov at a conference at Speyer in 1861:

> Starting from the assumption that each chemical atom possesses only a definite and limited amount of chemical force (affinity) with which it takes part in forming a compound, I might call this chemical arrangement, or the type and manner of the mutual binding of the atoms in a compound substance, by the name of 'chemical structure'.

He went on to add, using 'rational formula' in the sense of Kekulé's 'constitutional formula':

> Only one rational formula is possible for each compound, and when the general laws governing the dependence of chemical properties on chemical structure have been derived, this formula will express all of these properties.

This was a decisive advance from the position of Gerhardt, who was prepared to admit many formulae for one substance; they merely represented different kinds of reactivity.

What this meant in practice was that the 'typical' formulae would now be invested with a new significance: they represented how the atoms were connected to each other, not merely how they seemed to react. And from now on it was a case of one formula: one compound. For some years the old-fashioned type formulae continued to be used, though in this extended sense. But they clearly had their limitations, particularly for organic structures of any complexity. New types of formulae were urgently needed and several strange suggestions appeared in the early 1860s. Those that came eventually to be adopted for general use were embodiments of a daring verbal innovation by Edward Frankland. In 1866 he introduced into chemistry the term 'bond':

> By the term *bond* I intend merely to give a more concrete expression to what has received various names from different chemists, such as atomicity, an atomic power, and an equivalence.

But this was more than a new name; it was a move towards a very mechanical picture of a molecule. At first Frankland was inclined to think of intramolecular forces in gravitational terms but later he considered 'the bonds, which actually hold the constituents together being, as regards their nature, entirely unknown'. But for all that it now became possible to see the molecule in much more 'concrete' terms than ever before and this possibility gained expression in the 'graphic formulae' that were just now becoming popular.

2.4.3 Graphic formulae

Graphic formulae were devices to indicate something about a molecule rather more than what atoms were present. The use of simple parentheses as in $Al_2(SO_4)_3$ had a history going back to pre-structural days; it is a convenient way of indicating groups of atoms that tend to persist through many reactions. In

Figure 2.24 A. M. Butlerov (1828–86), Russian chemist at Kazan (1860–8) and St Petersburg (1868–80). After a European tour he produced the theory of structure in 1861, his most notable achievement. Thereafter he worked out the consequences of that theory in many researches in organic chemistry, including studies on isomerism and isomeric change, tertiary alcohols and the polymerization of formaldehyde. Other interests included, we are told, the administration of his family estate, spiritualism and beekeeping.

organic chemistry it was particularly favoured by the adherents of the radical theory. Their rivals, the 'typists', were far more anxious to indicate similarities between different classes of compound and found brackets convenient for this purpose. This is how Gerhardt indicated his four 'types' and their simple ethyl derivatives:

water type	*hydrogen chloride type*	*ammonia type*	*hydrogen type*

$$O\begin{cases}H\\H\end{cases} \qquad \begin{cases}H\\Cl\end{cases} \qquad N\begin{cases}H\\H\\H\end{cases} \qquad \begin{cases}H\\H\end{cases}$$

$$O\begin{cases}C_2H_5\\H\end{cases} \qquad \begin{cases}C_2H_5\\Cl\end{cases} \qquad N\begin{cases}C_2H_5\\H\\H\end{cases} \qquad \begin{cases}C_2H_5\\H\end{cases}$$

ethanol	chloroethane	ethylamine	ethane

These formulae *for Gerhardt* were devoid of structural significance but that did not prevent their being taken over by the early protagonists of valency, and this kind of representation continued long into the structural era. The addition of the marsh gas type by Kekulé, sometimes written as

$$C\begin{cases}H\\H\\H\\H\end{cases}$$

was a further powerful extension of this method. Yet for complicated organic molecules one type alone was insufficient, so Kekulé introduced the idea of *mixed types*. In his 1861 *Lehrbuch* he represented the lower carboxylic acids in terms of both hydrogen and water types:

types	*formic acid*	*acetic acid*	*propionic acid*

$$\begin{cases}H\\H\end{cases} \quad \left.\begin{matrix}H\\H\end{matrix}\right\}O \qquad \begin{cases}H\\CO\\H\end{cases}\!\!\!\Big\}O \qquad \begin{cases}CH_3\\CO\\H\end{cases}\!\!\!\Big\}O \qquad \begin{cases}C_2H_5\\CO\\H\end{cases}\!\!\!\Big\}O$$

hydrogen   water

This can happen when a 'polyatomic radical' (polyvalent radical) in place of two or three atoms of hydrogen causes 'the holding of these molecules together'.

Apart from the general messiness of such formulae what disadvantage did they have?

First, there was a strong element of *artificiality*. The formulae had to be related to one or more of the existing five types. Many, like Kolbe years before, had thought the limitation of organic nature to a mere handful of types 'nothing but a mere artifice'; exercises of this sort only served to deepen that impression. Secondly, even given the five types, there was an element or *arbitrariness* as to which ones were chosen. Thus compare the formula for acetic acid above with the more usual alternative at the time, based on the water type alone,

$$\left.\begin{matrix}C_2H_3O\\[4pt]H\end{matrix}\right\}O$$

or with this abomination perpetrated by G. C. Foster in 1865:

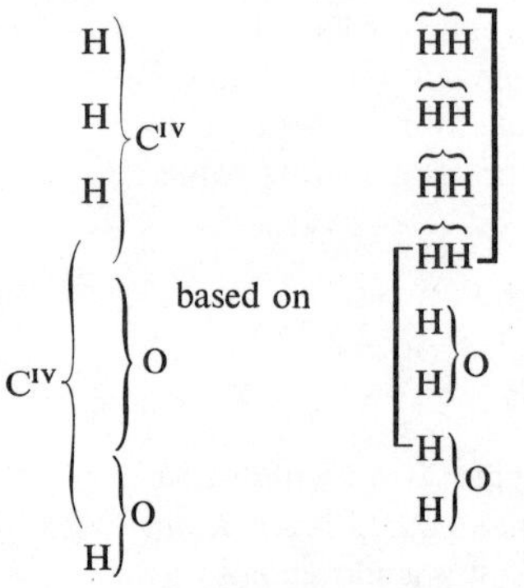

$$C^{IV}\Bigg\{ \begin{matrix}H\\H\\H\end{matrix}\Bigg\} C^{IV} \qquad \left[\begin{matrix}\widetilde{H}\widetilde{H}\\\widetilde{H}\widetilde{H}\\\widetilde{H}\widetilde{H}\\\widetilde{H}\widetilde{H}\end{matrix}\right]$$

based on

$$\left.\begin{matrix}H\\H\end{matrix}\right\}O \qquad \left.\begin{matrix}H\\H\end{matrix}\right\}O$$

This is a weird and wonderful combination of carbon, hydrogen and water types.

There was also a more serious disadvantage that became apparent later, and that was that such methods are totally inapplicable to cyclic compounds, as well as to many open-chain ones of considerable complexity. So various alternatives were tried, none of which need detain us long. Almost unknown were the touching and intersecting circles of Joseph Loschmidt (1861):

H   C   O   methanol   formic acid   acetylene

More famous were Kekulé's 'roll and sausage' formulae in which atoms are reckoned to be linked to others vertically above or below them.

marsh gas   ethane   acetic acid

SAQ 12 (*Objective 10*) To appreciate the strength and weakness of these formulae write down in these terms the two propanols:

$$CH_3CH_2CH_2OH \quad and \quad CH_3CH(OH)CH_3$$

You will notice that we have explained our dilemma by the familiar device of indicating bonds by lines. This stratagem was in fact born out of just this kind of situation. To demonstrate to yourself how great an advance this was, spend a few minutes on the following exercise.

SAQ 13 (*Objective 10*) Rewrite the following formulae in terms of bonds but do not add any extra information you may happen to have about the compounds:

(a) Kekulé's formula for sulphuric acid:

$$\begin{matrix} H \\ \\ SO_2 \\ \\ H \end{matrix} \begin{matrix} O \\ \\ \\ \\ O \end{matrix}$$

(b) Gerhardt's formulae for ethylamine and ethane.

(c) Kekulé's formulae for formic and propionic acids.

(d) Foster's formula for acetic acid.

(e) Loschmidt's formulae for methanol and formic acid.

The use of lines in formulae dates back to William Higgins in 1789, but they then had no structural significance. The first person to use them to represent valencies was Couper, though his formulae had certain defects which he was not able to rectify before ill-health compelled him to leave chemistry. While he was still at Edinburgh another Scotsman, Alexander Crum Brown, was preparing a thesis for the degree of MD 'On the Theory of Chemical Combination'. This rather un-medical title was quite typical at a time when a Scottish medical degree was one of the few British university routes to a career in chemistry. Whether Crum Brown and Couper met during those few months in 1858 is uncertain, but the remarkable fact about the thesis is that it gave to chemistry its first clear examples of graphical notation, showing bonds as clearly as if they were tie-bars in a piece of machinery. In 1864 his formulae were published in a paper to the Royal Society of Edinburgh. Atoms were represented by their chemical symbols enclosed in a circle; valency bonds were indicated by lines, at first broken but later continuous. The following examples show molecules

Figure 2.25 Alexander Crum Brown (1838–1922), Scottish chemist who read medicine at Edinburgh and chemistry at Heidelberg and Marburg. His MD thesis (1861) contains one of the first uses of graphic formulae and later publications by him and Frankland led to their widespread adoption. His studies included research on aromatic substitution and electrolyses of half-esters of dibasic acids. He also wrote on physiology and topology, maintaining a lifelong interest in three-dimensional knitting.

of ethane, ethylene and the two propanols; note how the difficulties presented by the latter to Kekulé are circumvented and how the double bond of ethylene is indicated:

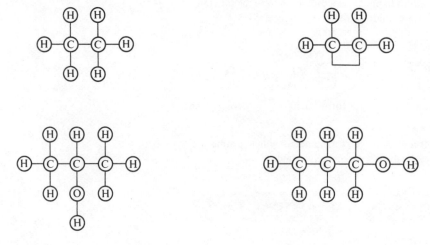

These new formulae, so totally committed to the theory of structure, had a mixed reception at first. Frankland wrote to Crum Brown in 1866:

> There is a good deal of opposition to your formulae here, but I am convinced that they are destined to introduce much more precision into our notions of chemical compounds. The water-type, after doing good service, is quite worn out.

Doubtless in that spirit, Frankland introduced these formulae into his *Lecture Notes for Chemical Students* that year, observing in the Preface:

> I have extensively adopted the graphic notation of Crum Brown which appears to me to possess several important advantages over that first proposed by Kekulé.

Some of Frankland's formulae are given below. Several points are noteworthy. They are not confined to organic chemistry, the valency of iron is regarded as a constant 6 so that ferric chloride is represented as Fe_2Cl_6 (in conformity with its vapour density) and unused valencies are indicated as satisfying each other as in $FeCl_2$:

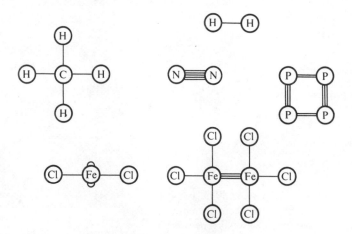

These formulae soon lost the circles around the letters and became those we are familiar with today. They made an immediate impact upon the audience for whom they were intended—the students in Britain's Mechanics' Institutes who were struggling to acquire a knowledge of science the hard way, by part-time study and often with little effective direct tuition. In such publications as *The English Mechanic*, available to them at 1d per week, graphic formulae like this may be encountered years before the academic chemists could rid themselves of their structural inhibitions sufficiently to make use of them. It is salutary to note, also, that the earliest discussions of valency were published in *The English Mechanic*, and the first occurrence of the word in print seems to have been in the issue for 12 March 1869. The leaders of chemical thought in England were still engaged in ideological disputes as to whether atoms really existed or not, but to less sophisticated minds the value of concepts like valency and structure were almost self-evident. Some indication of the confusion in high places may be gained from the following light-hearted jingle composed, in 1868, for a convivial gathering of certain Fellows of the Chemical Society who called themselves the 'B-Club'; the reference to Kay Shuttleworth is to the first Secretary to the Privy Council's Committee for Education:

'NH$_4$HSO$_4$ according to Frankland'

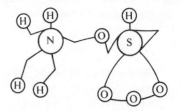

Figure 2.26 A contemporary skit on Frankland's notation.

> Though Frankland's notation commands admiration,
> As something exceedingly clever,
> And Mr Kay Shuttleworth praises its subtle worth,
> I give it up sadly for ever:
> Its brackets and braces, and dashes and spaces,
> And letters decreased and augmented
> Are grimly suggestive of lunes to make restive
> A chemical printer demented.
>
> I've tried hard, but vainly, to realise plainly
> Those bonds of atomic connexion,
> Which Crum Brown's clear vision discerns with precision
> Projecting in every direction.
>
> ... In fact, I am dazed with the systems upraised
> By each master of chemical knowledge,
> Who seems to suppose that truth only grows
> In the shadow of one little college.

We have already noted how Frankland's graphic formulae were applied by him to both organic and inorganic molecules, thus constituting a further example of the way in which the theory of structure, though originating in organic chemistry, was able to draw the two branches together in the 1860s. It was not long before quite complicated mineral substances were formulated in this way, as in the following example:

In fact this is all quite wrong and the long-term effects were losses rather than gains, but the immediate impact was undoubtedly to help forwards the celebrated cause of chemical unity.

2.4.4 Determination of structure: first moves

Until the 1860s structures were not known, though they might have been guessed. From now on there was a vast expenditure of effort to discover the structures of as many substances as possible, much, though not all of which, was to the enrichment of organic chemistry. During the early years several basic misconceptions were cleared up. Were there, for instance, two kinds of valencies associated with a saturated carbon atom? At one stage it had looked as though there were; but this was shown to be a result of experimental error.

Sometimes there was evidence for non-equivalence of all the affinities or valencies of other atoms, the classic case being the nitrogen atom in compounds like NH$_4$Cl. Butlerov pointed out the non-existence of NH$_5$ and NCl$_5$ and proposed—rightly—that not all five 'valencies' were equivalent.

In modern terms why is this?

Because nitrogen, in NH_4Cl, for example, is attached to four atoms, the fifth (chlorine) being ionic and therefore united to the ammonium ion electrovalently.

But were the *four* N—H valencies the same? That this was the case was shown conclusively by Hofmann in a somewhat similar way. The belief that ammonium salts might be addition compounds like $NH_3.HCl$ was disproved by Victor Meyer and Lecco in 1876 by obtaining the *same* product from methyl iodide and diethylmethylamine and from ethyl iodide and dimethylethylamine, so rendering highly improbable such formulations as $NMe_2Et.EtI$, and implying all four alkyl groups were equivalently placed:

$$NMeEt_2 + MeI \longrightarrow NMe_2Et_2I \longleftarrow NMe_2Et + EtI$$

Then, of course, came the nature of the benzene nucleus and the famous suggestions of Kekulé and many others. We cannot discuss it here, but some reference will be made later in the Course. With these and other basic problems cleared up, the way was open to tackle the structures of the thousands of natural and synthetic products available. For many years the classic approaches of syntheses and degradation led to fruitful results. The application of physical methods came later, and will be touched on in Unit 3.

SAQ 14 (*Objectives 11 and 13*) A famous book, *A Treatise on Chemistry* (1878) by Roscoe and Schorlemmer includes, not untypically for its time, the following formulae for inorganic compounds:

$$\left.\begin{matrix} Cl \\ Cl \end{matrix}\right\}O \qquad \left.\begin{matrix} I \\ I \\ I \end{matrix}\right\}As \qquad \left.\begin{matrix} F \\ F \\ F \\ F \end{matrix}\right\}Si \qquad O{\bigvee\!\!\atop O}\!\!O{\atop S}$$

65854—6

$$Cl-S-O-Cl \qquad \begin{matrix} HO \\ \\ HO \end{matrix}\!\!\!\searrow\!\!P-O-OH \qquad \begin{matrix} HO \\ \\ HO \end{matrix}\!\!\!\searrow\!\!HO-P=O$$

How far do these illustrate the thesis that the theory of structure in organic chemistry had a unifying influence on chemistry as a whole?

2.4.5 Into three dimensions

The extension of the structure theory into three dimensions was the almost simultaneous achievement of le Bel and van't Hoff independently in 1874. Before this time, others (most notably Louis Pasteur) had speculated about a possible three-dimensional arrangement for atoms in molecules. Pasteur had long been interested in the chirality of crystals, e.g. those of sodium ammonium tartrate which he had been able to separate into two enantiomorphous forms having equal but opposite effects on plane-polarized light. As far back as the 1840s he had speculated that the chirality of the crystal might be a reflection of a molecular chirality, suggesting in 1860 that an irregular tetrahedron might be the basis of the molecule. In the same paper he had drawn attention to a 'profound distinction between the products formed under the influence of life and all others': only a living organism could, he thought, produce true chirality. Here was a striking difference between the world of living organisms and the laboratory! True asymmetric synthesis was not to come under laboratory conditions until the present century. Pasteur's distinction can be read as that between living organisms and laboratories as sources of materials, rather than between organic and inorganic chemistry *per se*. In any event, so fast was the tide of unification running against him that he was not able to stem it even if he had wanted to. But the influence of Pasteur lingered on in Paris, and his views on molecular dissymmetry were the starting-point for the arguments of le Bel in 1874; a molecule with four different groups attached to the corners of a tetrahedron would be asymmetric

Figure 2.27 J. A. le Bel (1847–1930), French chemist working as an independent consultant. He is famous for his announcement, in 1874 and independently of van't Hoff, that optical activity in organic compounds is associated with the presence of an asymmetric carbon atom, and that their disposition in space is probably tetrahedral. He gave no diagrams and his work has been largely overshadowed by that of van't Hoff whom he had met while assisting Wurtz in Paris. He published little else of interest.

and optically active, but the asymmetry and the optical activity would vanish if any two of these groups became the same.

When van't Hoff proposed his tetrahedral carbon atom it was, he confessed, 'a continuation of Kekulé's law of the quadrivalence of carbon', and may also have owed something to the latter's models of carbon atoms, used for demonstration purposes in Bonn where van't Hoff had spent part of 1872.

The consequences for organic chemistry of the tetrahedral hypothesis were incalculable. Some of these will be the subjects of the TV programmes on stereochemistry. They include the explanation of optical and geometrical isomerism, the rise and fall of Baeyer's strain theory and the evolution of conformational analysis. For inorganic chemistry the results came much later; the first resolution of a compound with an asymmetric nitrogen atom was not until 1899. In so far as the birth of stereochemistry was a logical conclusion to the development of the theories of structure and valency it may be seen in general terms as a mild unifying factor. But, by 1874, divergent tendencies were becoming manifest again and stereochemistry was to play no significant part in uniting the two branches until Werner's resolution of an inorganic compound in 1911. That this had but a slight effect is a striking indication of the fragmentation of chemistry that preceded its unification in our own day. The forces at work in chemistry over the last hundred or so years will be the subject of the next Unit.

Figure 2.28 J. H. van't Hoff (1852–1911), Dutch chemist known, with le Bel, as discoverer of the tetrahedral carbon atom and also as one of the founders of physical chemistry. After studying with Kekulé and then Wurtz he became a lecturer at the Veterinary College at Utrecht (1876), later succeeding to the Chairs in Amsterdam and Berlin. His research in physical chemistry included the theory of solutions, reaction kinetics, solid solutions, thermodynamics, phase rule, etc. He became the first Nobel Prizewinner in Chemistry, in 1901.

SAQ answers and comments

SAQ 1 1 In extent: Wöhler's 'synthesis' was one transformation, but Berthelot spent 10 years on a whole series of reactions.

2 In motivation: Wöhler's discovery was accidental, but Berthelot was determined to eliminate vitalism.

3 In starting materials: Wöhler began with a substance as complex as his product—in fact isomeric with it—and one of dubious inorganic character; Berthelot commenced with very simple substances, in some cases the elements themselves.

4 In product: Wöhler obtained urea in excellent yield; Berthelot's syntheses often gave tiny quantities of the substance required.

5 In claim: Berthelot's claim that syntheses, his among them, could 'obliterate all the demarcation lines between mineral and organic chemistry' is in marked contrast to Wöhler's more modest rejoinder (Unit 1, Section 1.2.1.3).

Finally, as we shall go on to show, they differed:

6 In significance: the significance of Wöhler's work of 1828 was chiefly for the light it threw on isomerism; of Berthelot it must be said that, if his syntheses did have an effect on eliminating vitalism, then they were by no means the only or even most important cause; their chief value seems to have been in helping to establish the idea of general methods in organic chemistry.

SAQ 2 *Positively*: the isolation of what were believed to be 'radicals' might have been expected to give a new boost to the radical theory, much as Bunsen's work with cacodyl had done. In fact, as we have seen, it came too late.

Negatively: the formation of diethylzinc was a further blow to any lingering belief in uncompromising dualism. The electronegative iodine in ethyl iodide was replaced by the electropositive zinc. The absence of outcry comparable with that occasioned by the discovery of hydrogen replacement by chlorine can only reinforce the impression that old-fashioned dualism had lost its power and charm.

SAQ 3 Both were young men, each determined to make a name for himself. Perkin wished to produce a substance of great practical utility, the anti-malarial drug quinine. Frankland wanted to be the first to isolate an entity of great theoretical importance, the radical 'ethyl'. Each set himself an impossible task, but each made valuable discoveries en route (coal-tar dyes and organometallic compounds). Neither was guided by a theory of structure. Perkin, however, was led to his experiments by consideration of molecular formulae. Frankland went a little further and took into account relative reactivities, e.g. the affinity between potassium or zinc and iodine. He was, moreover, much more strongly directed by his commitment to the theory of radicals—a good example of how a theory that is partially correct can lead to excellent experimental research. Finally, note how both men, having tried one substance unsuccessfully, replaced it by a simpler one: allyltoluidine by aniline and ethyl cyanide by the iodide. This indicates a recognition of the 'relatedness' between compounds that is characteristic of a mature chemical approach.

SAQ 4 (a) The contemporary view was generally that (a) = 4 *or* (a) = 1, since doubts were felt as to the *organic* nature of urea and ammonium cyanate. For defeat of the vitalist cause it would have been necessary for (a) = 2 or (a) = 3. This underlines one of the difficulties in answering a question of this kind, namely the confusion between a product of a living organism on the one hand and its decomposition products on the other.

(b) Taking the sequence on p. 10 the first step is 5 (since CS_2 was not believed to occur in nature); the next three are all 4; only the last one is 2. Here again there was room for debate because in nature acetic acid occurs directly as a result of decomposition and corruption as in vinegar. There was no general belief that micro-organisms might be responsible for initiating such processes as the souring of wine.

(c) Given urea as generally organic this must be 2. Otherwise it is simply 4. The absence of excitement following Davy's discovery was partly because he failed to recognize the product.

(d) This is a degradation from a complex to a more simple molecule and is therefore not a synthesis in any of the senses above.

(e) This was 4 since butane was not then regarded as a natural product.

(f) In view of the origin of benzaldehyde (f) = 6.

(g) Clearly (g) = 4.

(h) As acetylene occurred previously only through degradation of organic matter (e.g. in coal-gas) (h) = 5.

(i) Methane, however, as marsh gas, was a natural product, though again a result of degradation, like urea. According as to whether or not it was accorded full organic status (i) = 2, or, more probably, (i) = 4.

(j) Another case of 4.

(k) Since both reagents are found in nature (k) = 1.

(l) Again (l) = 1.

SAQ 5 (a) On ideological grounds there was a good deal of criticism of Berthelot's work because it was accomplished under conditions so very different from those in nature (e.g. at high temperatures or with electric sparks, etc.). In the end, indeed, Berthelot's ideology was almost inverted because his syntheses served to emphasize how *different* were the processes in the laboratory and in the living organism.

(b) In Berthelot's syntheses many of the individual products were obtained in very low yield, partly because of the complexity of the final mixtures. Such drastic treatments would be expected to produce all kinds of odd reactions and, in the absence of highly specific catalysts, would not today be reckoned a likely preparative tool of any value. For demonstration of a theoretical point this may not have mattered, but in this respect the syntheses stand in stark contrast with those of industrial Germany where chemistry and economics were becoming increasingly intertwined.

SAQ 6 From these data the molecular weight of carbon disulphide will be $2 \times 38 = 76$. Of these 76 'atomic weight units' 15.8 per cent will be due to carbon, i.e. 12. In a similar way the figures for the remaining examples are 12, 12, 24, 12, 24, 12, 24, 24, 72. Hence *in those cases studied* the minimum value for carbon is 12, and this is taken as its atomic weight.

There are two assumptions:

1 That Avogadro's hypothesis is true.

2 That the samples taken were adequately representative of carbon compounds. In time, of course, thousands of such measurements have been made and no one has ever found evidence for less than 12 units due to carbon. If one day 6 were found a great deal of rethinking would be necessary!

42

SAQ 7 (a) The parts by weight of oxygen combining with (say) 100 parts of manganese are thus:

I: 28.1   II: 42.17   III: 56.2

and these are in the ratio 2:3:4. Hence we could have MnO_2, MnO_3 and MnO_4 or MnO, Mn_2O_3 and MnO_2. Since II is isomorphous with Fe_2O_3 it must be Mn_2O_3 and I is MnO and III is MnO_2. The atomic weight for manganese follows from any of these. In I, for example, 28 parts of oxygen combine with 100 parts of manganese, so, since there is one atom of each, the atomic weight is

$$\frac{16}{28.1} \times 100 = 56.92$$

(b) If 177.26 parts by weight of nitrogen combine with 100 of oxygen, then 16 of oxygen combine with (16×1.7726) parts of nitrogen, i.e. 28.37. Assuming the volumes of reacting gases are proportional to the numbers of *atoms* it follows that there are twice as many atoms of nitrogen as of oxygen. Hence the empirical formula for the combined product would be N_2O and the atomic weight of nitrogen $(\frac{1}{2} \times 28.37) = 14.13$.

You will note that both these figures are close to the modern ones (especially when corrected from $O = 16$), though in case (b) for the wrong reasons. Avogadro would have provided a surer path to the same answer.

SAQ 8 (a) Dumas now believes that 'the assimilation of the two chemistries becomes more and more near'.

(b) The distinction between empirical knowledge and the understanding of the nature of the forces that act; also the distinction between organic substances and 'organized matter' (i.e. living organisms).

(c) That they may be composed of even simpler matter: 'the elements of mineral chemistry may be complex'. He arrives at this conclusion by analogy from organic compounds as well as on the basis of atomic weight differences. The series Li, Na, K is of course significant, but the one Mg, Cu, Fe is not.

SAQ 9 (a) Very few; you may have identified the two oxides of carbon (25 and 28) and three oxides of nitrogen (23, 26 and 27) together with SO_3 (31).

(b) The major pre-Karlsruhe advances were:

1 Use of alphabetical symbolism introduced by Berzelius in 1814.

2 Abandonment of clusters of atoms with an apparently structural significance (see also (d) below).

3 Recognition of 7, 11, 12 and 18 as compounds.

4 Partial acceptance of combining volumes of gases leading to, among other things, correct formulae for water (21) and methane (25). Hence also:

5 At the hands of Berzelius especially, correct atomic weights for more elements and hence correct molecular formulae for more compounds.

(c) By about 1864 there was a nearly universal acceptance of atomic weights approximating to those in use today. Hence the uncertainty over many molecular formulae was removed. Also the controversy over the meaning of gaseous volumes came to a speedy end, so enabling one to be sure (for instance) that ethylene was C_2H_4 and not CH_2 (or even CH as in 24).

(d) The one thing that was still lacking was any clear conception of structure. In respect of that deficiency the clusters of Dalton were not significantly improved upon until well after Karlsruhe. They seem to have been formulated from considerations of symmetry and neatness, rather than from any other motive; there was no possible way, for instance, for Dalton to know that ◯●◯ was better than ●◯◯ for carbon dioxide. Nor, in 1860, had any substantially new chemical reasons emerged. But there was an awareness that such formulae might be dangerous and suggest more than they were meant to. Their very abandonment was a sign that something like a doctrine of structure was struggling for articulation, however, and when in a few years this was formulated it enabled chemists to see just how far they had travelled since Dalton. These events are the theme for the remainder of this Unit.

SAQ 10 If we set aside questions relating to personal circumstances we can identify several areas of possible misunderstanding. One must always assume misunderstanding rather than deliberate misrepresentation unless circumstances point the other way, and in this case they do not. Let us assume, then, that each man was convinced of the rightness of his own claim, though aware of the assertions made by the other. Here are some of the chemical issues that could have clouded the issue.

1 The multiple meanings of 'valency': as we have seen there are at least four legitimate meanings that can be given to that term in the 1860s. The idea of catenation was clearly due to Kekulé and Couper, and Frankland could claim nothing here; the issue of the tetravalent carbon atom was more complex, but Frankland did not take the step of saying that a polyvalent atom could hold together the rest of the molecule; however, it is when we consider the saturation capacity of the elements that we find a real difficulty, for Frankland certainly seems to have asserted this in 1852 and Kekulé just as vigorously denies his claim. We therefore have to explore further.

2 The use of different atomic weight values. It will be recalled that Frankland wrote before Karlsruhe, and before Kekulé. Their early contributions to valency were, respectively, on the bases C = 6, O = 8, and C = 12, O = 16. Yet would that be sufficient ground for Kekulé to reject Frankland's claim outright? Were correct atomic weights a prerequisite for the perception of valency? We return to this point in the text.

3 The different kinds of proposition. Kekulé, as we have pointed out (Section 2.3.2), proposed that atoms or groups of atoms could serve to unite the rest of the molecule. Frankland's proposition was a general 'law' about 'symmetrical grouping of atoms' for which he offered no hypothesis. In fact Kekulé did attack his rival on precisely those grounds.

4 The different chemical traditions. One thing is abundantly clear: Frankland approached valency from radicals, Kekulé from types. If this seems to us a purely trivial matter, certainly one not big enough to cause deep misunderstandings, even estrangements, then this is because we have failed to enter deeply enough into the spirit of 19th century chemistry. Again, we return to this ideological divide later.

SAQ 11 Because it implies that one compound may have two structures and its molecule is therefore not uniquely defined by one. A classic case in the 1860s was acetoacetic ester, assigned an enolic formula by Geuther and a ketonic one by Frankland and Duppa. It was soon realized that these corresponded to two quite distinct molecular species existing in dynamic equilibrium with each other:

$$CH_3.C(OH):CH.COOC_2H_5 \qquad CH_3.CO.CH_2COOC_2H_5$$
$$\text{enol} \qquad\qquad\qquad\qquad \text{keto}$$

However, for some while this kind of behaviour caused considerable difficulties to those advocating a theory of structure.

SAQ 12 The case of propan-1-ol is straightforward and you should have obtained a formula something like this:

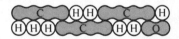

However, with propan-2-ol there are complications, and it is impossible to arrange the atoms in two layers as above. Instead something like the following results:

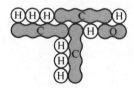

This means that the three hydrogen atoms at the bottom must be joined horizontally, not vertically, but there is nothing in the formula itself to show that this must be so. The reason for the complication is that the basic skeleton is branched, whereas in propan-1-ol it is not:

$$
\begin{array}{cc}
C-C-C & C-C-O \\
| \quad\text{or} & | \qquad \text{as opposed to } C-C-C-O \\
O & C
\end{array}
$$

Similar complications arise for any branched-chain molecule.

SAQ 13

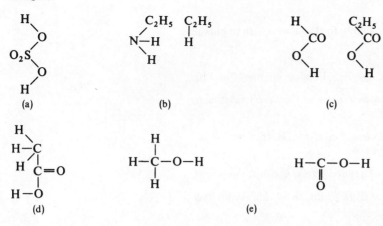

(a)   (b)   (c)

(d)   (e)

SAQ 14 The first three formulae illustrate a direct takeover by inorganic chemistry of the ideas and notation of the theory of types, so effectively extending that theory to all other non-metals. The next two formulae (for thionyl chloride and sulphur trioxide) are attempts to conserve a valency of 2 for sulphur, and reflect the preoccupation of Kekulé and other organic chemists with constant valency. But the last formulae, alternative representations for orthophosphoric acid, are highly significant. The first of them conserves the 3-valency of phosphorus (and is incorrect, suggesting a peroxide grouping), but the second has conceded a valency of 5. The reason for this concession could only have been (and was) the production of this acid 'by decomposing PCl_5 or PF_5 with water', i.e.

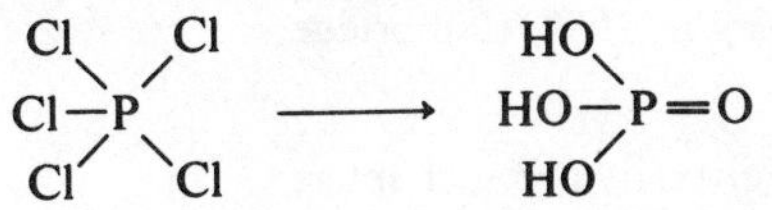

and *this* is the kind of argument based on synthesis that was becoming so common in organic chemistry. Clearly such ideas were being applied in the same way to both branches. It is significant that one of the authors, Schorlemmer, occupied the Chair of Organic Chemistry at Manchester and that the treatise was on chemistry *as a whole*.

Further reading

The following suggestions, as for Unit 1, are simply for those who wish to pursue matters further; first some general reading:

John Read (1957) *Through Alchemy to Chemistry*, Bell, London, chapters 9 and 10.

A. Findlay and T. I. Williams (1965) *A Hundred Years of Chemistry*, Methuen, London, chapter 3.

A. J. Ihde (1966) *The Development of Modern Chemistry*, Harper and Row, New York, Evanston and London, chapters 6–8.

T. H. Levere (1971) *Affinity and Matter*, Clarendon Press, Oxford, chapter 6.

J. E. Jorpes (1966) *Jac. Berzelius, his Life and Work* (trans. B. Steele), Almqvist & Wiksell, Stockholm.

J. R. Partington (1964) *A History of Chemistry*, vol. 4, Macmillan, London.

J. H. Brooke (1973) Chlorine substitution and the future of organic chemistry, *Stud. Hist. Phil. Sci.*, **4**, 47–94.

C. A. Russell (1968) Berzelius and the development of the atomic theory, in D. S. L. Cardwell (ed.) *John Dalton and the Progress of Science*, Manchester University Press, pp. 259–73.

In addition, the following bear more specifically on the subject matter of Unit 2:

W. E. Palmer (1965) *A History of the Concept of Valency to 1930*, Cambridge University Press, chapters 1–3.

C. A. Russell (1971) *The History of Valency*, Leicester University Press, chapters 1–6.

Harold Hartley (1971) *Studies in the History of Chemistry*, Clarendon Press, Oxford, chapters 3, 5, 8 and 9.

W. H. Lehmann (1972) *Atomic and Molecular Structure: the Development of our Concepts*, Wiley, New York, chapters 2–4.

R. F. Gould (ed.) (1966) *Kekulé Centennial*, American Chemical Society, chapters 1 and 2.

Acknowledgements

Grateful acknowledgement is made to the following for material used in this Unit:

Figure 2.3 Bulletin de la Société Chimique de France, 1913, **13**, 1; *Figures 2.4, 2.5, 2.8, 2.10, 2.11, 2.12, 2.17, 2.19, 2.21, 2.25 and 2.28* Chemical Society Library Portrait Collection; *Figure 2.6* Picture Archive, Philipps University, Marburg; *Figure 2.7* Southampton University Library; *Figure 2.9* North Western Museum of Science and Industry, Manchester; *Figures 2.13 and 2.16* Stadtbibliothek, Karlsruhe; *Figure 2.24* Richard Anschutz (1929) *August Kekulé*, Verlag Chemie Gmbh, Weinheim/Bergstr; *Figure 2.27* C. A. Bischoff (1894) *Handbuch der Stereochemie*.

Unit 3   Specialism and its hazards

Contents

3.0   Introduction

3.0.1   Scope of the Unit

The last Unit was confined very largely to the short period of time—10 or 15 years—following 1860 in which the two branches of chemistry showed such a high degree of convergence that it looked as though the cry of 'one chemistry' was at last to be justified. The culmination of Berthelot's work in organic synthesis coincided almost perfectly in time with the calling of the Karlsruhe Conference that was to lead, though in a very odd way, to the virtually universal acceptance of one system of atomic weights. The emergence of a clearly articulated and clearly illustrated theory of valency served as another unifying factor, capable of application to organic and inorganic compounds alike. And the theory of structure derived from valency became the common property of all chemists. About a hundred years ago it was possible to look back on two decades in which the science had been completely transformed. Today, a chemist who has occasion to refer to earlier literature will usually be acutely aware of the much greater difficulty in comprehending papers written before 1860 compared with those from the 1870s onwards. From then on, we are in a real sense in the modern era.

However, this highly successful clearing-up operation did not lead, as many confidently supposed it would, to a unified science at all. Instead there followed another divergent phase, much longer than any before it, in which the optimistic welcomes to 'but one chemistry' were seen to be decidedly premature, to say the least. Not only did the inorganic and organic chemists go their separate ways, the situation was further complicated by the emergence of a third contender, physical chemistry. So instead of one changing interface there now became three: organic/inorganic, organic/physical and inorganic/physical. Contrary to what one might have expected, perhaps, the traditional structures now became hardened and the newcomer, instead of healing the breach, actually helped to extend it.

These developments form the subject matter for the first part of this Unit and they take us up to about the period of the Second World War. From just after the war until the present the prophets of unity have had a good time. There now appears to be a much greater degree of unification in chemistry than was ever experienced in the 19th century. The causes and marks of this tendency will be discussed in the second part of the Unit, although he would be rash who suggested that chemistry has settled into anything like a permanent shape. Finally we shall bring the Unit, and this Block, to a close by some reflections on what is meant by assertions about processes of unification in science and what the implications may be for scientific progress in general.

We are therefore going to be concerned with more than a century of chemistry and this may seem a disproportionate length of period for such a brief treatment, certainly in comparison with the previous Unit, which covered barely 20 per cent of that time. In that connection, there are two things that need to be borne in mind. The first is that history does not necessarily proceed at an even pace, and events of great significance may occur closely packed in time (as in the 1860s) or spread out much further. As we shall see, the latter seems to have been the case from the 1880s to 1940s so far as the structure of chemistry is concerned, though that is not to assert that trends of great importance did not develop *within* the three main branches. And, of course, in other areas of science the most portentous events were taking place, whether in evolutionary biology, nuclear physics or general relativity. The second point is that many of the themes touched on in this Unit will be returned to again and again in the rest of the Course, so that in a sense we are merely introducing them here. You should therefore regard this material as a link between Units 1 and 2 and those that follow, as well as being important in its own right.

Table A

List of chemical terms, concepts and principles in Unit 3

Introduced in a previous Unit	Unit Section No.	Developed in this Unit	Page No.
	S100[1]		
catalysis	12.4.2	acetoacetic ester	20
electron pair	8.4.4	adsorption	31
flame spectroscopy	6.4.4	Brownian motion	36
hydrogen bond	10.5.2	ferrocene	38
ideal gas equation	5.2	law of mass action	27
lanthanides	8.2	mesomerism	26
osmotic pressure	13 (App. 2)	osazones	22
periodic law	8.2	Ostwald's dilution law	29
reaction kinetics	11.4	tautomerism	20
	S24–[2]		
1 : 4 addition	8.1		
Grignard reagent	7.2		
infrared spectroscopy	1		
organometallics	7.2.2		
ultraviolet spectroscopy	1.4		
	S25–[3]		
coordinate link	7.4		
coordination number	3.4		
dipole moment	1.2.3		
effective atomic number	8.6.5		
molecular orbital theory	8.3		
third law of thermodynamics	4.10		
	ST 294[4]		
electrolytic dissociation	E1.2.3		
third law of thermodynamics	T3.4.2		

[1] The Open University (1971) S100
Science: A Foundation Course,
The Open University Press.

[2] The Open University (1973) S24–
An Introduction to the Chemistry of Carbon Compounds,
The Open University Press.

[3] The Open University (1973) S25–
Structure, Bonding and the Periodic Law,
The Open University Press.

[4] The Open University (1975) ST 294
Principles of Chemical Processes,
The Open University Press.

3.0.2 Objectives

At the end of this Unit you should be able to:

1 Evaluate the role of (a) research output, (b) periodical publication and (c) professional recognition, as indicators of the fragmentation (or otherwise) of chemistry.
(SAQs 1, 15)

2 Identify the disruptive effects of valency theory on the unity of chemistry from about 1870 to 1900.
(SAQs 2, 3, 4, 5)

3 Identify the distinctive features of inorganic chemistry from 1870–1900 which tended to isolate it from organic chemistry.
(SAQ 8)

4 Characterize the main features of Werner's view of inorganic complexes.
(SAQs 6, 7)

5 Recognize growth-inhibiting features of inorganic chemistry between the two World Wars.
(SAQ 9)

6 Identify the chief features of organic chemistry from about 1870 to the 1930s.
(SAQs 10, 11)

7 Evaluate the role of physical chemistry in uniting the two other branches.
(SAQ 12)

8 Evaluate the thesis that the structure of chemistry is unaffected by events in the external world.
(SAQ 13)

4

9   Recognize the main features of chemistry in the last twenty-five years that have led to much closer integration than before.
(SAQ 15)

10   Identify the changing role of organometallic chemistry in the unification of chemical science.
(SAQ 14)

3.0.3   Related course material

TV programme 3:   This is an introduction to the TV course on stereo-chemistry and concerns the origins and development of conformational analysis. It is much more closely related to the following television programmes than it is to the material in this Unit although the following connections should be borne in mind:
(a) the utility of the tetrahedral carbon atom model;
(b) the importance of instrumentation;
(c) the changing attitude of organic chemists to physical chemistry.

3.0.4   Some external indicators*

As we enter a period that may reasonably be called 'modern' we find a number of new phenomena in the organization of chemistry that can act as 'indicators' of the extent to which it is, or is not, in a phase of convergence towards unification. These are mainly to do with the emergence in the later 19th century of a creature known as a 'professional chemist'. In addition, there are several more familiar yardsticks of a rough-and-ready sort that can always be applied to gain some kind of measure of such trends. The most obvious of these is the research output of individuals.

3.0.4.1   Research outputs of individuals

Much work is currently in progress in identifying research and other chemical interests of many individuals in the last and present centuries. Much also remains to be done. While it is too early to report precise analyses, a few general impressions are clear and do not seem likely to be superseded. From the 1870s onwards there is a marked tendency for individuals to concentrate their research in one of the main fields: organic, inorganic or (later) physical chemistry. This is abundantly obvious for the people who became leaders of chemical thought, i.e. those with the highest research output. The case of Hofmann we have already met; Kekulé was similar in concentrating almost entirely in the organic area. So were Kolbe, Perkin, Baeyer, Fischer and a host of others. There were far fewer workers whose output was predominantly in the region of inorganic chemistry; Werner is the chief example, though he began research on organic compounds. The new science of physical chemistry was to attract more devotees in the late 19th century, among them being van't Hoff, Arrhenius, Helmholtz, Ramsay, Nernst and Ostwald. It is often hard to classify a paper as 'physical' or 'inorganic' during this period, and this evidence seems to suggest that, although the boundary between organic chemistry and the rest had hardened, that between inorganic and physical chemistry was much more fluid. Of course there were individuals, such as Berthelot, who contrived to subvert the historians' neat classification and worked in all three fields at roughly the same time, but they were fairly exceptional. On the whole, this kind of evidence does suggest that at least until the last war organic chemistry continued to diverge from the rest, while in the early part of the present century the inorganic/physical interface tended to harden, though to a rather smaller extent.

* In this Section many titles and dates appear. There is no suggestion that these should be learnt; they are for background information only.

The growth of chemical literature in the period following 1870 offers a further indication of the structure of chemistry as it was seen at the time. Certainly there is no lack of material to study. The 19th century saw an enormous output of textbooks, most of which were either devoted to one of the three divisions or else attempted a comprehensive survey of chemistry but divided their material into these three categories after a general introduction. It was also the time for the appearance of the specialist journal. Chemistry was to have several of these to itself, one of the earliest being the French *Annales de Chimie* of 1789 and, after several changes of title, still surviving. Equally prestigious have been Liebig's *Annalen der Chemie und Pharmacie* from 1832 and the *Journal für praktische Chemie* of 1834, later to be edited by Kolbe. In Britain, chemists had to share the pages of more generalist scientific journals such as the *Philosophical Magazine* and the Royal Society's *Philosophical Transactions*. Britain's first exclusively chemical publication lasted precisely one year, being largely ignored by all except the humblest class of chemical student: *The Chemist*, 1824–5. A similar short lifespan awaited *The Laboratory* which, by paying its contributors, denied itself the privilege of survival within a few months of its birth in 1867. The resources of the Chemical Society ensured the continuance to this day of its *Journal*, founded in 1848. The French analogue was the *Bulletin de la Société chimique de France* of 1858 and was followed ten years later by its German counterpart the world-famous *Berichte*, while the *Journal of the American Chemical Society* began in 1879. Meanwhile, less formal periodicals became viable propositions, of which a noted example was the *Chemical News* founded by William Crookes in 1860.

Thus it is obvious that science itself was finding it necessary to move towards specialist periodicals within the 19th century, and this trend has continued ever since. But what about the specialization *within* chemistry? It is here that an inspection of journal titles reveals some interesting trends. Thus consider the founding dates of the following journals:

> *Journal of Physical Chemistry*, 1896
> *Journal of Organic Chemistry*, 1936
> *Inorganic Chemistry*, 1962

At face value this seems to suggest that chemistry became specialized in the order (1) physical, (2) organic, (3) inorganic. But in the previous Units we have seen that exactly the reverse was the case.

SAQ 1 (*Objective 1*) What important factors are ignored in this false deduction?

One cannot ignore the prior existence of generalist journals which would have included papers on these specialisms. The point about the appearance of a new specialist periodical is that it occurs, not when the subdiscipline first becomes recognized, but when there is no room for it in existing journals. Hence it is at least partly a function of editorial policy of those earlier journals. This becomes immediately apparent when one compares the date of *Inorganic Chemistry* (1962) with that of the German *Zeitschrift für anorganische Chemie* (1892). Clearly, circumstances in the United States and Germany were totally different. In the former the modest developments in chemistry as a whole could well be accommodated within the generous pages of the existing *Journal of the American Chemical Society* or of its analytical or industrial counterparts. But in Germany the growth-rate in chemistry was so much greater that new journals simply had to appear. Moreover, an inspection of all except the earliest issues of the journals edited by Liebig and Kolbe reveals their overwhelming preoccupation with organic chemistry, which in itself is sufficient explanation for the absence of a separate organic journal in that country. For many decades after 1860 it was very nearly the case that, in Germany, chemistry itself *was* organic. In America, physical chemistry was more prominent than inorganic (thanks largely to the efforts of J. W. Gibbs). But even there the Germans were ahead and the world's first journal in that area was the *Zeitschrift für physikalische Chemie* of 1887.

So it does not make sense to use the first appearance of a specialist journal as an indicator of the subject's novelty or otherwise. One might as well conclude that no one recognized the practice of teaching chemistry before the foundation of the *Journal of Chemical Education* in 1924! All that founding dates show with certainty is that, by then, the subject was well recognized in its own right. Fortunately, however, better indicators are to hand in analyses of the contents of general chemical periodicals and in certain of the non-periodical publications. These imply that the latter half of the 19th century witnessed such a concentration on organic research, especially but by no means exclusively in Germany, that its divergence and separateness are self-evident. The supreme monument of this tendency is to be found in the mammoth *Handbuch der organischen Chemie* of F. K. Beilstein, commenced in 1880 and still (under the auspices of the German Chemical Society) being updated. Its 54 huge volumes covering the literature till 1919 are a permanent reminder of the days when organic chemistry was the single pursuit of the majority of chemists and a specialism in its own undoubted right.

Fig. 3.1 Friedrich Konrad Beilstein (1838–1906), German chemist who became Professor at St. Petersburg in 1866. He worked in both organic and inorganic chemistry, but is chiefly known for his monumental *Handbuch* (see Figure 3.2), begun while he was still an assistant to Wöhler in Göttingen and first published in the early 1880s.

Figure 3.2 Beilstein's *Handbuch der organischen Chemie* in the OU Library at Walton Hall. This enormous work is intended to include a brief reference to every organic compound known, classified into three groups (acyclic, homocyclic and heterocyclic) and arranged according to their molecular formulae. The first three editions were the work of Beilstein himself, but all later production has been handled by the German Chemical Society. The problem, of course, is the time-lag between a compound's discovery and its first appearance in *Beilstein*. The fourth edition consisted of 27 volumes appearing between 1916 and 1937, covering the literature up to 1910. Another 27 volumes constituted the first supplement and related to the years 1910–19. This photograph shows the basic (4th) edition, together with the first, second and third supplements and part of the fourth.

3.0.4.3 Professional recognition

The 19th century was the time when chemistry ceased to be just a subject and became a profession as well. It might therefore be expected that a divergent tendency within it might be recognizable by official job-titles for those within one or other of its diverging parts. In other words, when did people start calling themselves 'organic chemists', 'physical chemists' or even 'inorganic chemists'? To take just the British case, we can ask when university chairs were created for each of the specialisms. But again we have to be on our guard, for other factors may be present. Thus at Cambridge the first Chair in Physical Chemistry was established in 1920 but the one for Organic Chemistry did not appear until after the Second World War. Yet the real beginnings at Cambridge go back to 1875 with the appointment of one of Hofmann's former pupils, Ruhemann, as Demonstrator in Organic Chemistry. A violent quarrel with Dewar seems to have been the cause of his subsequent eclipse, so it is certainly unwise to infer too much about the structure of chemistry from this.

Other evidence accords with the view that the 1870s marked a beginning for the professional organic chemist, at least in England. In 1874 the first British Chair in Organic Chemistry was established in Manchester, Schorlemmer being the first incumbent.* That university did not appoint a Professor in Physical Chemistry, however, until 1933. Chairs in Organic Chemistry at University College, London (1902), Leeds (1904) and Kings College, London (1905) confirm the impression that by the turn of the century the subject was well recognized in Britain as a specialism capable of independent existence. And both here and abroad the term 'organic chemist' and 'organic laboratory' had passed into common speech.

---

* Berthelot was appointed to a chair of Organic Chemistry at the École de France in 1865.

3.1   Specialization—PHASE V

We have now seen some of the evidence that prompts the belief that organic chemistry after 1870 had moved so much from the parent subject that it could be treated in practice as a virtually independent branch of science. In that sense chemistry had certainly been divergent.

It is now necessary to enquire into some of the causes of this movement of specialization; they fall into two groups, those that are inherent in the growth of chemistry itself ('internal') and those that are not, which we can designate 'external'. In the following account, the first four factors fall into the 'internal' category, the fifth being a collection of influences coming largely from outside chemistry:

1   Controversy over valency.

2   The growth of specialist inorganic chemistry.

3   The growth of specialist organic chemistry.

4   The emergence of physical chemistry.

5   External factors.

3.1.1   Controversy over valency

The last 30 years of the 19th century were marked by a general erosion in the confidence about valency that had seemed so strong in the 1860s. It is debatable how far this reflected, or how far it helped to cause, the increasing scepticism about atoms themselves. Certainly the two trends went hand-in-hand. Extreme philosophical positions adopted by chemists like Berthelot were not conducive to progress in structural theories generally. However, for valency, the chemical issues were nearly all centred round one critical problem: that of variable valency. Could an atom ever display more than one valency?

The assertion that valency was a fixed parameter was frequently made in the 1860s, the most articulate and persistent spokesman of that point of view being Kekulé: 'atomicity [valency] is a fundamental property of the atom which should be as constant and invariable as the atomic weight itself'. There can be no doubt as to his reasons. He considered that his own distinctive contribution to valency theory had been the recognition of just such a specific and invariable property of the atom. On the other hand he regarded Frankland as having missed the essential point by confusing atoms and equivalents. But there was more to it than his own personal reputation. In Kekulé's hands valency had emerged from a study of *organic* chemistry, and this is the one area where a fixity of valencies appears to be shown: carbon 4, hydrogen 1, oxygen 2, nitrogen 3, etc. But when this basically organic concept is applied to inorganic chemistry one soon runs into trouble. The net result was a divergence of the two branches. It was an inversion of the situation in Phase II when the powerful application of the Berzelius analogy to inorganic chemistry led to much useful discovery (see Unit 1). Now, as we shall go on to show, that same analogy was to suffer such an overload when forced to carry the concepts of fixed valency that the system largely broke down. Consequently the last third of the century (and much of the present one) saw a strong divergence.

It would not be fair to give the impression that even organic chemistry was completely at home with fixed valency. In the areas of unsaturated and aromatic compounds there were problems. By about 1870 most chemists accepted that in ethylene, for instance, the two carbon atoms were linked by a double bond, though a minority view that survived into our own century was that each carbon atom was linked to three other atoms and so its valency must be 3, the molecules being written:

$$
\begin{array}{ccc}
H & & H \\
\diagdown & & \diagup \\
& C-C & \\
\diagup & & \diagdown \\
H & & H
\end{array}
$$

The problem as to why a double bond should confer reactivity, and so appear as 'weaker' than a single bond, was partly met in Baeyer's strain theory (see TV programme 3) and partly in Thiele's theory of partial valencies (see Section 3.1.3.2). Similar considerations applied in the realm of aromatic compounds, but time does not permit us to examine them here. By and large, however, it was true that organic chemists could manage quite well with fixed valencies and a minimum of fuss about possible exceptions.

For inorganic compounds the problems were far greater. Some examples follow in the next SAQ.

SAQ 2 (*Objective 2*) The following structures were suggested during the period. What principles are their originators trying to preserve? Suggest any ways in which evidence could have been raised against them *at that time*.

(a) $H_3N=NH_3$ for ammonia

(b) $O-N\equiv N-O$ for nitric oxide

(c) $Cl-Hg-Hg-Cl$ for mercurous chloride

(d) $C_6H_5-O-S-O-O-H$ for benzenesulphonic acid

(e)
$$\begin{array}{c} Cl \quad\quad Cl-Cl-K \\ \diagdown \quad \diagup \\ Pt \\ \diagup \quad \diagdown \\ Cl \quad\quad Cl-Cl-K \end{array}$$
for K_2PtCl_6 (from $PtCl_4 + KCl$)

These examples represent two common stratagems to maintain constant valency: doubling up formulae and, more frequently still, arranging atoms in long chains. The latter practice may be seen as a further 'carry-over' of organic procedures into the other realm; examples like these offer an obvious analogy with organic catenation:

phosphorus pentoxide: $O=P-O-O-O-P=O$
sulphuric acid: $H-O-O-S-O-O-H$
potassium trithionate: $K-O-O-O-S-S-S-O-O-O-K$

(In the last example we now know that there *are* S—S links, though not O—O links.)

More effective than these devices was the one proposed by Frankland in which 'latent atomicities' on atoms could unite with each other, thereby lowering the 'active atomicity' or effective valency by two units at a time. An example was his representation of the two chlorides of iron:

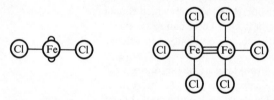

SAQ 3 (*Objective 2*) What objections could be made against such practice? Bear in mind that many new compounds of transition elements were being discovered at this time.

We have seen that the hypothesis that an element always displayed the same valency was too simple-minded for all the facts, particularly those of inorganic chemistry. Quite different from the approaches described above was the fall-back position to which Kekulé and others were driven. This was the assumption of 'molecular compounds'. As early as 1864 Kekulé drew attention to 'attraction... felt even between atoms which are found to belong to different molecules'. In certain circumstances, two molecules may 'adhere together, so to speak, thus forming a group endowed with a certain stability, always less strong than an atomic combination, however'. This was a *molecular compound*. Examples, to illustrate the fixed trivalency of nitrogen and phosphorus and the monovalency of iodine include:

	NH_3, HCl	PCl_3, Cl_2	ICl, Cl_2
for	NH_4Cl	PCl_5	ICl_3

Such compounds would dissociate very easily, as on vaporization. This was indeed the case for ammonium salts and others examined, though later work denied that this always happened, as with phosphorus pentafluoride, PF_5.

Chemical arguments soon became available, of which just one must suffice. The quaternary ammonium compounds were imagined to be molecular compounds in order to avoid the embarrassment of pentavalent nitrogen, this element being then widely regarded as trivalent.

SAQ 4 (*Objective 2 and revision*) What chemical argument could have been advanced to dispel the molecular compound view of quaternary ammonium compounds?

Despite setbacks of this kind the molecular compound idea lingered on until the 20th century. Because of its lack of specificity it turned out to be a useful dumping-ground for awkward cases. This was especially true in organic chemistry, where the classic problem of substances like naphthalene picrate (an addition product of naphthalene and picric acid) resisted all attempts at explanation until the advent of quantum theory. But the most interesting case of all is that which we should call today the *hydrogen bond*. Here, if anywhere, something of the situation envisaged by Kekulé still obtains; discrete molecules are linked together by weak forces which can be overcome at higher temperatures. The reactions of organic and inorganic chemists to this phenomenon were entirely characteristic. Thus it was discovered from the vapour density of acetic acid just above its boiling temperature that the molecules were associated in pairs, while in the 1880s studies of hydrogen fluoride vapour revealed extensive polymerization. The former passed almost unnoticed, or, if brief reference was made to it, the matter was passed off by chemists in the organic tradition as a loose combination, rather like a hydrate. On the other hand, inorganic chemists speculated on whether the fluorine might not be able to combine with itself, and in the 1920s they were cheerfully canvassing the possibilities of hydrogen atoms with a co-valency maximum of two. Modern conceptions of the hydrogen bond tend to resemble the molecular compound view, with partly electrostatic forces linking the positively polarized (and small) atom of hydrogen with negatively polarized atoms of oxygen, fluorine etc.:

-------- = hydrogen bond

The inorganic opposition to molecular compounds was associated with a general antipathy to Kekulé's constant valency. The American chemist Remsen, presenting to his students the formula $PtCl_4 . 2KCl$, observed 'That period [punctuation dot] has for many years been a full stop to thought. Don't let such devices keep you from trying to find out what lies behind them'. In similar vein Reitzenstein derided a molecular compound as 'a concept in which all kinds of things could be included without thereby achieving the slightest clarification'. And perhaps the most noted inorganic chemist of the last century, Mendeléev, complained that 'Kekulé's division of chemical compounds into "atomic" and "molecular" types is artificial, arbitrary and unsound'. The final emancipation of inorganic chemistry from such bondage came at around the turn of the century with the work of Werner (Section 3.1.2.2). One result of this, as we shall see, was a further deepening of the divide between the two traditional branches of chemistry.

SAQ 5 (*Objective 2 and revision*) How far was Kekulé's advocacy of molecular compounds and fixed valencies a reflection of earlier ideas in chemistry?

Figure 3.3 Group at meeting of the British Association in Manchester in 1887.

Back row (*left to right*):

J. A. Wislicenus (1835–1902), German organic chemist, who succeeded Kolbe at Leibzig. He developed the synthetic uses of acetoacetic ester and other compounds, promoted the general acceptance of the tetrahedral carbon atom and made notable contributions to the study of geometrical isomerism.

G. Quincke (1834–1924), German colloid chemist, Professor of Physics at Heidelberg from 1875 until 1907.

H. E. Schunck (1820–1903), Manchester calico-printer who, after studying at Giessen, developed an extensive research programme in natural dyestuffs.

C. Schorlemmer (1834–92), German chemist, assistant (1859) and Professor of Organic Chemistry (1874) at Owens College, Manchester. His writings include a collaborative *Treatise on Chemistry*, with Roscoe, and several works on the history of chemistry.

J. P. Joule (1818–89), English physicist who had studied under Dalton, and later collaborated with William Thomson in research in thermodynamics. Of independent means, he worked for most of his life in Manchester. His most famous paper, describing experiments to establish 'The mechanical equivalent of heat', was read in 1847.

Front row (*left to right*):

Lothar Meyer (see Figure 2.16).

D. I. Mendeléev (1834–1907), Russian chemist famous for his role in the enunciation of the periodic law and the formulation of the Periodic Table. The uniqueness of that role has been, and is, a matter of some controversy, for others, including Lothar Meyer, were independently coming to similar conclusions. Mendeléev's contribution was massive, however, and included the prediction of several 'missing' elements that were subsequently discovered. From 1861 to 1890 he taught at St Petersburg, where he was a colleague of Beilstein and Butlerov. A man of fiery temperament and unorthodox appearance, he was remarkable (at that time) for permitting himself but one haircut a year.

H. E. Roscoe (1833–1915), English chemist who studied under Graham and Bunsen and then succeeded Frankland at Owens College, Manchester. He worked in photochemistry and was the first to isolate pure vanadium. His *Treatise*, in collaboration with Schorlemmer, was oustandingly successful, as was his advocacy of, and contributions to, popular education in science. From 1885 to 1895 he was Liberal MP for South Manchester, and then Vice Chancellor of London University till 1902.

Figure 3.3 Group at meeting of the British Association in Manchester in 1887.

Back row (*left to right*):

J. A. Wislicenus (1835–1902), German organic chemist, who succeeded Kolbe at Leibzig. He developed the synthetic uses of acetoacetic ester and other compounds, promoted the general acceptance of the tetrahedral carbon atom and made notable contributions to the study of geometrical isomerism.

G. Quincke (1834–1924), German colloid chemist, Professor of Physics at Heidelberg from 1875 until 1907.

H. E. Schunck (1820–1903), Manchester calico-printer who, after studying at Giessen, developed an extensive research programme in natural dyestuffs.

C. Schorlemmer (1834–92), German chemist, assistant (1859) and Professor of Organic Chemistry (1874) at Owens College, Manchester. His writings include a collaborative *Treatise on Chemistry*, with Roscoe, and several works on the history of chemistry.

J. P. Joule (1818–89), English physicist who had studied under Dalton, and later collaborated with William Thomson in research in thermodynamics. Of independent means, he worked for most of his life in Manchester. His most famous paper, describing experiments to establish 'The mechanical equivalent of heat', was read in 1847.

Front row (*left to right*):

Lothar Meyer (see Figure 2.19)

D. I. Mendeléev (1834–1907), Russian chemist famous for his role in the enunciation of the periodic law and the formulation of the Periodic Table. The uniqueness of that role has been, and is, a matter of some controversy, for others, including Lothar Meyer, were independently coming to similar conclusions. Mendeléev's contribution was massive, however, and included the prediction of several 'missing' elements that were subsequently discovered. From 1861 to 1890 he taught at St Petersburg, where he was a colleague of Beilstein and Butlerov. A man of fiery temperament and unorthodox appearance, he was remarkable (at that time) for permitting himself but one haircut a year.

H. E. Roscoe (1833–1915), English chemist who studied under Graham and Bunsen and then succeeded Frankland at Owens College, Manchester. He worked in photochemistry and was the first to isolate pure vanadium. His *Treatise*, in collaboration with Schorlemmer, was oustandingly successful, as was his advocacy of, and contributions to, popular education in science. From 1885 to 1895 he was Liberal MP for South Manchester, and then Vice Chancellor of London University till 1902.

Figure 3.3 Group at meeting of the British Association in Manchester in 1887.

Back row (left to right):

J. A. Wislicenus (1835–1902), German organic chemist, who succeeded Kolbe at Leibzig. He developed the synthetic uses of acetoacetic ester and other compounds, promoted the general acceptance of the tetrahedral carbon atom and made notable contributions to the study of geometrical isomerism.

G. Quincke (1834–1924), German colloid chemist, Professor of Physics at Heidelberg from 1875 until 1907.

H. E. Schunck (1820–1903), Manchester calico-printer who, after studying at Giessen, developed an extensive research programme in natural dyestuffs.

C. Schorlemmer (1834–92), German chemist, assistant (1859) and Professor of Organic Chemistry (1874) at Owens College, Manchester. His writings include a collaborative *Treatise on Chemistry*, with Roscoe, and several works on the history of chemistry.

J. P. Joule (1818–89), English physicist who had studied under Dalton, and later collaborated with William Thomson in research in thermodynamics. Of independent means, he worked for most of his life in Manchester. His most famous paper, describing experiments to establish 'The mechanical equivalent of heat', was read in 1847.

Front row (left to right):

Lothar Meyer (see Figure 2.19)

D. I. Mendeléev (1834–1907), Russian chemist famous for his role in the enunciation of the periodic law and the formulation of the Periodic Table. The uniqueness of that role has been, and is, a matter of some controversy, for others, including Lothar Meyer, were independently coming to similar conclusions. Mendeléev's contribution was massive, however, and included the prediction of several 'missing' elements that were subsequently discovered. From 1861 to 1890 he taught at St Petersburg, where he was a colleague of Beilstein and Butlerov. A man of fiery temperament and unorthodox appearance, he was remarkable (at that time) for permitting himself but one haircut a year.

H. E. Roscoe (1833–1915), English chemist who studied under Graham and Bunsen and then succeeded Frankland at Owens College, Manchester. He worked in photochemistry and was the first to isolate pure vanadium. His *Treatise*, in collaboration with Schorlemmer, was oustandingly successful, as was his advocacy of, and contributions to, popular education in science. From 1885 to 1895 he was Liberal MP for South Manchester, and then Vice Chancellor of London University till 1902.

3.1.2 The growth of specialist inorganic chemistry

3.1.2.1 Elements and the Periodic Table

Many of the developments in inorganic chemistry in the last part of the 19th century were concerned, directly or indirectly, with the determination of atomic weights. Underlying this quest was something rather more fundamental than what has been said of the physics of that time, that it was simply 'the search for the next decimal place'. Magnificent research by Stas, Marignac and others was informed at least in part by a concern to know whether the then accepted elements really were 'simple', or whether the old hypothesis of Prout, that everything was made of hydrogen, might not in fact be true. As we saw in Unit 1, Section 1.3.2, some chemists drew further encouragement for this view from the non-simplicity of organic radicals. Given the possibility of Prout's being right, atomic weights should be integral multiples of the atomic weight of hydrogen. By mid-century it was becoming clear that this was not likely to be the case, but much useful quantitative work continued.

It must be at once obvious that this kind of research was very different from that of the organic chemist, and the kinds of experimental skills involved would have been very different. So, of course, would the theoretical constructs employed. Indeed one feature of work that is basically concerned with stoichiometry (combining quantities) is that very little theory is needed. Whereas the organic chemist was committed to atoms, the workers in stoichiometry could get along quite happily with equivalents. This was to be of great significance.

The concentration on atomic weights led to a gradual recognition of a periodic relationship between them. It was a combination of this fact with the increasing awareness of valency that led directly to the enunciation of the periodic law. Without implying that Mendeléev had sole rights to the title of its discoverer, we would do well to note his remark that 'the periodic law brings out the dependence of these two magnitudes, atomic weight and valency, on one another' (cf. S25–, Unit 6). It also provided for inorganic chemistry its first great generalization and gave it a framework comparable, perhaps, to the idea of homologous series in organic chemistry.

In the Periodic Table of 1872 Mendeléev indicated a number of vacant spaces for elements yet to be discovered. Thanks largely to this stimulus, three of these elements were found within the following fifteen years: gallium, scandium and germanium. The Periodic Table was also useful in checking valencies proposed

Reihen	Gruppe I. — R^2O	Gruppe II. — RO	Gruppe III. — R^2O^3	Gruppe IV. RH^4 RO^2	Gruppe V. RH^3 R^2O^5	Gruppe VI. RH^2 RO^3	Gruppe VII. RH R^2O^7	Gruppe VIII. — RO^4
1	H=1							
2	Li=7	Be=9,4	B=11	C=12	N=14	O=16	F=19	
3	Na=23	Mg=24	Al=27,3	Si=28	P=31	S=32	Cl=35,5	
4	K=39	Ca=40	—=44	Ti=48	V=51	Cr=52	Mn=55	Fe=56, Co=59, Ni=59, Cu=63.
5	(Cu=63)	Zn=65	—=68	—=72	As=75	Se=78	Br=80	
6	Rb=85	Sr=87	?Yt=88	Zr=90	Nb=94	Mo=96	—=100	Ru=104, Rh=104, Pd=106, Ag=108.
7	(Ag=108)	Cd=112	In=113	Sn=118	Sb=122	Te=125	J=127	
8	Cs=133	Ba=137	?Di=138	?Ce=140	—	—	—	— — —
9	(—)	—	—	—	—	—	—	
10	—	—	?Er=178	?La=180	Ta=182	W=184	—	Os=195, Ir=197, Pt=198, Au=199.
11	(Au=199)	Hg=200	Tl=204	Pb=207	Bi=208	—	—	
12	—	—	—	Th=231	—	U=240	—	— — —

Figure 3.4 Periodic Table proposed by Mendeléev in 1871 (*Annalen*, Supp. vol. viii, p. 151). This is one of the earliest versions in something approaching a modern form. Note the displacements of La and Ce in Group IV and the ambivalence of the positions of Cu, Ag and Au. Note also the gaps deliberately left, especially at 44, 68 and 72. What were the missing elements?

on other grounds. Many years later it was to prove the guiding principle behind the new electronic theory of valency. But it is all too easy to overstate its importance for suggesting lines of research. Its main value for the rest of the 19th century seems undoubtedly to have been in the training and education of chemists. Indeed, it is not going too far to say that the most important discoveries in inorganic chemistry for the rest of the century not only owed little to the Periodic Table but actually offered it an embarrassing challenge.

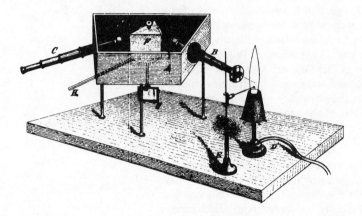

Figure 3.5 Spectroscope of Bunsen and Kirchoff (*Annalen der Physik*, 1860, **110**). This is, of course, a device for examining *emission spectra* of elements, these being obtained by heating the material in a flame and passing the light successively through the lens system in B, the prism F and the eyepiece system C. With this simple visual method several new elements (including Cs, Rb and Tl) were almost immediately discovered.

The early 1860s had seen a resumption in the discovery of new elements. Within a few months of the invention of the spectroscope by Bunsen and Kirchhoff in 1860 the use of this instrument had led to the discovery of caesium, rubidium, thallium and indium. These elements were shortly to find their homes in the Periodic Table, as was fluorine, isolated by Moissan in 1886. But in 1878 a new development began in the isolation of the first of a whole series of metals that appeared to require housing in the same box as lanthanum. By 1907, fourteen had been discovered: cerium, dysprosium, erbium, europium, gadolinium, holmium, lanthanum, lutecium, neodymium, praseodymium, samarium, terbium, thulium, ytterbium (cf. S25–, Unit 3). These are, of course, the lanthanides, but the discovery of such a new inner series removed some of the optimistic belief in the finality of the Periodic Table as then received.* Still more was this true of the discovery of the inert gases (Group 0) at the turn of the century. But their appearance, as will be mentioned later, paved the way for the Lewis' electronic theory of valency and new moves to unite chemistry once again.

See AST 281, TV programme 1.

In all this work one factor may easily go unnoticed. Almost all of the elements discovered and examined, with the exception of argon, are either extremely rare, or extremely reactive, or both. Nor, with the exception of the use of cerium in gas-mantles, did any of them have the slightest commercial value at that time. Consequently there was never more than a minute number of workers involved in their study, and inorganic chemistry as a subject for academic research was an exceedingly limited field. As for the impression conveyed to the undergraduate in the 1890s we have this recollection from Sir Harold Hartley:

> And so in the 'nineties when I and my contemporaries were learning chemistry, organic chemistry gave us a tidy, logical picture in which the agreement between properties and structure showed clearly that the model represented a close approximation to the truth. In physical chemistry we got a picture of the dependence of chemical reaction and equilibria on the kinetics of the individual molecule or ion, and of the value of thermodynamics in giving generalised laws independent of any theory. Inorganic chemistry was more difficult; by then there were so many isolated facts to be remembered and, while the Periodic Table was a great help, there were many anomalies still to be explained.

* And, of course, these lanthanides did not display the regular gradation in valencies found elsewhere. Many of them favoured a 3-valent state.

3.1.2.2 Werner and coordination chemistry

Within a few years, however, the face of inorganic chemistry was to change dramatically as much of the confusion lamented by Hartley was cleared up in the powerful new theories of Alfred Werner (1866–1919). Werner was Professor of Chemistry at Zurich, where he developed the important concepts that were to find full expression in his famous textbook of 1905: *Neuere Anschauungen auf den Gebiete der anorganischen Chemie*. Later Units in this Course will expound Werner's ideas more fully than is necessary at this point. But it will be useful to give a very short account of five of their salient features.

Figure 3.6 Alfred Werner (1866–1919), French chemist who may be justly acclaimed as the founder of coordination chemistry. Apart from a year with Berthelot in Paris his studies and later services to chemistry were all in Zurich, where he became full professor in 1895. His early work was in organic chemistry (including studies of the isomerism of oximes and cyclohexane rings), but from about 1899 he concentrated almost exclusively on bringing order into the confused realm of coordination compounds. The climax of his work was the resolution, with V. L. King, of the first optically active compound of this class (1911). Two years later he received the Nobel Prize for chemistry.

In the first place, Werner set his face against a concept of valency as a directed unit force. At an early stage he had asserted that 'affinity is an attractive force acting equally from the centre of the atom toward all parts of its spherical surface', though this was far from the 'totally new idea' that some Werner enthusiasts have assumed it to be. Many other writers, especially in Germany, had proposed substantially the same thing. For Werner the departure from the traditional position was made more urgent by his research on transition metal complexes which led to the blunt assertion: 'the valency of the elements is not constant',

Secondly, Werner introduced the concept of 'coordination number' and abandoned the old idea of 'molecular compounds'. When, for instance, platinic chloride accepted two molecules of ammonia he denied both the formulation $PtCl_4 . 2NH_3$ and the other devices to 'save' the 4-valency of platinum. Since two such ammines were known, Cleve had suggested they might be

$$\begin{matrix} Cl & H_3N-Cl \\ & \diagdown \; Pt \diagup \\ Cl & H_3N-Cl \end{matrix} \quad \text{and} \quad \begin{matrix} Cl \\ | \\ Cl-Pt-H_3N-H_3N-Cl \\ | \\ Cl \end{matrix}$$

But for Werner this obscured their strong similarity, and he proposed that all six ligands should be attached to the central platinum atom, which thus displayed a coordination number of 6:

$$\underset{\underset{Cl}{\overset{|}{\diagup}}{\underset{Cl}{\overset{Pt}{\diagdown}}Cl}}{\overset{H_3N}{\diagdown}\overset{Cl}{\overset{|}{\diagup}}NH_3}$$

It was obvious that no conventional view of valency could accommodate such suggestions. Accordingly, Werner had recourse to an idea that had been widely canvassed in the last quarter of the 19th century. This was the notion of 'residual affinities', or 'auxiliary valencies', a perfectly natural deduction from his denial of valency as a directed unit force. In $PtCl_4$ and in NH_3 why should there not be some of that force left over in the platinum and nitrogen atoms which could then unite them in the complex? Thus, representing auxiliary valencies by a dotted line,

$$Cl_4Pt \ldots + 2 \ldots NH_3 \longrightarrow Cl_4Pt \ldots (NH_3)_2$$

This use of auxiliary valencies is a third characteristic of Werner's position. A fourth one is the concept of 'zones of affinity', referred to in the following quotation of 1893 as 'spheres':

> If we think of the metal atom as a ball, the six complexes directly bound with the same are found in a sphere described about the latter, and the rest that are found beyond this first sphere lie in a second sphere. For all the compounds being considered, we can propound the general law: The valency of the radical formed by the metal atom and the six complexes bound with it in the first sphere is equal to the difference between the valency of the metal atom and that of the monovalent groups in the first sphere, and wholly independent of the molecules present in the first sphere as water, ammonia, *etc.*

Thus ligands could be deemed to exist in one of two spheres, the maximum number in the inner zone being the coordination number, which was often 6. Groups in the outer sphere were less tightly bound to the central atom and were thus ionizable. Hence a test for chlorine in the outer sphere was the formation of a silver chloride precipitate on treatment with silver nitrate solution. Similarly, conductivity measurements could indicate the number of ionic species present. Thus a compound $Co(NH_3)_5Cl_3$ precipitated two equivalents of AgCl and had a conductivity compatible with the presence of three ions. It was thus

$$[Co(NH_3)_5Cl]Cl_2 \quad \text{or} \quad [Co(NH_3)_5Cl]^{2+} \ 2Cl^-.$$

In order to avoid implications about the equivalence or non-equivalence of ligands within the first sphere, structures were written non-committally thus:

$$\left[Co \begin{matrix} (NH_3)_4 \\ \\ (NO_2)_2 \end{matrix} \right]^+ NO_2^-$$

SAQ 6 (*Objective 4*) Apply Werner's rule concerning the overall valency of a radical in the quotation above to determine its value for each of the following atomic groupings (the valency of cobalt is 3):

(a) $[Co(NH_3)_6]$

(b) $\left[Co \begin{matrix} (NH_3)_3 \\ \\ (NO_2)_3 \end{matrix} \right]$

(c) $\left[Co \begin{matrix} (NH_3)_2 \\ \\ (NO_2)_4 \end{matrix} \right]$

A final mark of Werner's general theory is his extension to inorganic compounds of stereochemistry. Assuming that a 6-coordinate metal would have an octahedral arrangement, and a 4-valent metal a planar or a tetrahedral one, he was able to predict optical isomerism in cobalt complexes not containing an asymmetric carbon atom. In 1911 his assistant, Victor King, became the first to resolve

Figure 3.7 Victor L. King (1886–1958), American chemist who, as Werner's research student, became the first person to resolve an inorganic complex deriving its optical activity *only* from the asymmetry of the 6-coordinate metal (and not from any asymmetric carbon atoms in ligands). In June, 1911, he resolved *cis*-chloroamminebis (ethylenediamine)cobalt(III) chloride, by fractional crystallization of its salt with (−)bromocamphorsulphonic acid.

such a complex, so demonstrating to the world the essential correctness of Werner's beliefs concerning 6-coordinate complexes and their geometry:

en = $H_2N-CH_2-CH_2-NH_2$; the two N atoms of each ligand occupy two of the six sites round the Co atom. Verify for yourself the absence of plane or centre of symmetry.

It has been often said that in these researches Werner laid the foundations of modern inorganic chemistry.* The question that faces us now is how far they contributed to the unification of chemistry as a whole, and how far they tended to retard it. It certainly appears that they helped to emphasize the link between organic and inorganic chemistry. Werner himself had a distinguished record in organic research before he moved over to the other branch in the 1890s. He was not slow to draw upon his experience in one area to illumine the other. He resumed the practice of analogical argument, pointing out, for example, how quaternary ammonium salts can be regarded in a similar way to his inorganic complexes:

In fact he concluded that the organic chemist's theory of structure was only a special case of his own coordination theory.

SAQ 7 (*Objective 4*) How could Werner hold to this view?

Consistent with this belief Werner was able to claim that 'the difference still existing between carbon compounds and purely inorganic compounds disappears' as a result of his successful resolution of the cobalt complexes. Of this achievement one recent author (G. B. Kauffman) has suggested:

> The last brick in the crumbling wall of separation between inorganic and organic chemistry had been razed. The demolition begun eighty-six years earlier by Friedrich Wöhler with his artificial synthesis of urea from ammonium cyanate had been completed by Alfred Werner.

SAQ 8 (*Objective 3 and revision*) What features of this statement do you find (a) convincing, (b) unconvincing?

* These points are further developed in Unit 13.

17

3.1.2.3 The decline of inorganic chemistry

In order to avoid being carried away by exaggerated assessments of the immediate impact of Werner's work we need to bear in mind two things about inorganic chemistry just before the First World War. The first, as we suggested in connection with SAQ 8, is that it was simply not true that coordination complexes played a key role in inorganic chemistry either then or for 40 years ahead. What Werner did do in his own fairly short lifetime was to convince people that *in this area that posed so many questions* his theory was a satisfactory explanation. Its relevance to the chemistry of the free elements, the hydrides, the solid state, the phenomena of radioactivity and so on had yet to be perceived. Certainly he applied his ideas to non-transitional elements, as in the hydration of sulphur trioxide, but as alternative explanations were readily available for this kind of reaction, his argument was less impressive:

$$O=\overset{\overset{\displaystyle O}{\|}}{\underset{\underset{\displaystyle O}{\|}}{S}}\cdots + \cdots OH_2 \longrightarrow O=\overset{\overset{\displaystyle O}{\|}}{\underset{\underset{\displaystyle O}{\|}}{S}}\cdots OH_2 \longrightarrow O=\overset{\overset{\displaystyle OH}{|}}{\underset{\underset{\displaystyle O}{\|}}{S}}-OH$$

The second point is that represented by Sidgwick in his *Electronic Theory of Valency* of 1927:

> To the ordinary chemist it appeared that Werner had not succeeded in dethroning structural chemistry so far as organic compounds were concerned, but that he had provided a new theory of structure which was of great value in accounting for the behaviour of inorganic compounds, especially of those complex or 'molecular' compounds with which the old structural theory had proved itself incompetent to deal. This was clearly an unsatisfactory position; it represented molecules as being built on two different and apparently irreconcilable plans, and it emphasised an obviously false distinction between organic and inorganic chemistry.

So much for the unification of chemistry! Yet Sidgwick went on to say that the (to him) false distinction between the two branches 'must ultimately be resolved', and this he believed had happened in his own electronic interpretation of co-ordination, first published in 1923. This was based upon the concept of the shared electron-pair advanced by Lewis in 1916 (and itself much indebted to Werner), together with the octet theory of Lewis and Langmuir. As applied to first row elements of the Periodic Table the Sidgwick theory was able to account for the ability of beryllium and boron fluorides to accept further fluoride ions. Electron pairs donated by these ions formed dative covalent bonds, e.g.

$$:\!\overset{\cdot\cdot}{F}\!: \overset{\cdot\cdot}{Be}\, :\!\overset{\cdot\cdot}{F}\!: + 2 :\!\overset{\cdot\cdot}{F}\!:^- \longrightarrow :\!\overset{\cdot\cdot}{F}\!: \overset{\overset{\textstyle :\overset{\cdot\cdot}{F}:}{}}{Be} :\!\overset{\cdot\cdot}{F}\!:^{2-}$$

or $$BeF_2 + 2F^- \longrightarrow BeF_4^{2-}$$

Since the coordination number of first-row elements was 4 no further ligands could be added to the BeF_4^{2-}, or to BF_4^- or NH_4^+. But the behaviour of the second-row elements suggests that the coordination number for these must be 6, to take into account such ions as AlF_6^{3-} and SiF_6^{2-}. The important point in all this is the formation of an electron pair bond by donation of both electrons from the same atom—the 'dative bond' or 'coordinate link'. This view, allied with his own concept of effective atomic number (further discussed in Unit 13), led Sidgwick to a comprehensive view which, he said, 'breaks down the supposed distinction between organic and inorganic chemistry'. That this was, in principle, a further step towards unification cannot be doubted but it did not go as far as he claimed. In practice, its effects in this direction were not large until after the Second World War. One reason was undoubtedly the lack of a sound physical basis for the electron pair theory. Why, for instance, should pairing of electrons be so essential? Why should coordination numbers rise from the first to second to third rows in the Periodic Table? Thus BF_4^- is in accord with the octet rule, but SiF_6^{2-} is not (a kind of paradox the organic chemists did not have to face). Why (for that matter) should sharing of electrons in any manner constitute a chemical bond? The answers to these and other questions began to emerge with the application of quantum theory and wave mechanics in the late 1920s and 1930s, though at first the mathematical difficulties were dauntingly formidable.

Figure 3.8 N. V. Sidgwick (1874–1952), English chemist who, having taken a first in natural sciences at Oxford, went back and read classics (to please his family) obtaining another first two years later. Thereafter his devotion was to chemistry. He studied under Ostwald before returning to Oxford, where he spent the rest of his life, becoming Professor in 1935. His outstanding contribution was to apply the concepts and results of physical chemistry to the materials of organic and inorganic chemistry. *The Electronic Theory of Valency* (1927) is a magnificent synthesis of new and old ideas and exerted a lasting influence on chemical thought.

A further reason for the delayed integration of inorganic chemistry with the organic branch was that, despite Werner and Sidgwick, there was for such a long time so little to integrate. Nyholm commented on the British experience:

> From the first world war onwards inorganic chemistry languished, particularly in this country. There were a few schools where the study of complex compounds continued, and some important work was done, but as the workers were limited in the number of techniques which were of value to them, their conclusions as to structure were, in most cases, only tentative. Excellent preparative work was carried out on various subjects, such as the boron hydrides, but in comparison with organic and physical chemistry the amount of such work was small.

SAQ 9 (*Objective 5*) What is Nyholm's point about techniques? Which ones are meant?

We have already heard Sidgwick's optimistic evaluation of the electronic theory's role in unifying chemistry in 1927. Twenty-three years later his *Chemical Elements and their Compounds* appeared at the threshhold of a new era of unification. But as he looked back to the previous three decades he saw a further reason for slow progress. During that time inorganic chemistry has been 'usually so overburdened with the details of mineralogy, metallurgy, technical chemistry, and analysis that hardly any space is left for the consideration of the theoretical relations'. It might be added that it was usual for inorganic textbooks to devote considerable space to what was sometimes known as 'General Chemistry'. This would include radioactivity, atomic structure and the role of the electron in valency. The inherent interest of such topics must have made the contrast still more pointed with much of what followed. But when one writer (A. W. Stewart, 1931) wrote of 'the tremendous revolutions in inorganic chemistry and radioactivity which have been brought about by the work of Thomson, Ramsay, Rutherford, Soddy, Aston and Lewis', his list of names identifies the area of progress as that of 'General Chemistry' almost exclusively.

Thus the picture that emerges of inorganic chemistry as a coherent body of theory from about the beginning of the Franco–German war in 1870 to the end of the Second World War is one of rapid progress under Mendeléev and Werner, followed by a period of relative decline. Nothing happens during this time to effect a radical realignment with the chemistry of carbon compounds. As the *Annual Reports of the Chemical Society* gloomily observed in 1947:

> Inorganic chemistry is slowly—too slowly perhaps—changing from a descriptive and preparative science to one concerned with valency, structure and reaction mechanisms.

3.1.3 The growth of specialist organic chemistry

We last took our leave of organic chemistry at a time when it was widely and loudly proclaimed that 'there is but one chemistry'; vitalism had been banished, valency had emerged as a key to inorganic and organic compounds alike, the theory of structure had been established ostensibly in the organic field but in principle applicable throughout all chemistry, and a genuine spirit of ecumenical harmony was fostered by a nearly universal acceptance of one system of atomic weights. With such a legacy from the 1860s and early 1870s it is almost a pity to have to record that thereafter, for something like 75 years, organic chemists pursued their studies with increasing independence and self-sufficiency. We have already noted that their inorganic counterparts were unable to achieve a satisfactory and coherent view of their subject, despite the twin advances of the Periodic Table and the coordination theory of Werner. So it is perhaps not surprising that the organic chemists were content to leave them to it. The deeper they went into their own subject the more aware they became that they had problems enough and to spare and for which their own resources seemed sufficient without drawing upon the Wernerian or other concepts of inorganic chemistry.

Several factors contributed to the continued separation of organic chemistry. Some we have already met in connection with controversies over valency. Others will be encountered in the next two Sections. Now, however, we must note those that were strictly internal to the subject itself.

3.1.3.1 Synthetic methods

Not for the first time we encounter organic syntheses as important determinants of the structure of chemistry. The fact is that nothing succeeds like success. Given the great versatility of the synthetic methods developed in the last few decades of the 19th century it is hardly surprising that they should have been exploited with such vigour and enthusiasm. Behind the virtual explosion of organic synthesis in this period lay various external pressures (see Section 3.1.5) but other factors included the recognition of the *generality* of many organic syntheses, as well as the fact that they could be conducted with a *maximum economy of theory* yet with a reasonable certainty as to the *structure of the products*. The output of synthetic compounds was so prodigious, especially in Germany, that it would be unreasonable to expect that all this work (or even most of it) led to significant advances in the understanding of chemistry. As Stewart observed in 1918:

> It is safe to say that the latter half of the nineteenth century will be regarded as a time when theoretical speculation played the main part in the development of the subject. Of the hundred thousand organic compounds prepared during that time, the majority were still-born and their epitaphs are inscribed in Beilstein's Handbook.

In 1931 the same writer, who was Professor at Queen's University, Belfast, while welcoming the advent of various 'interesting' classes of compounds observed that:

> It must be admitted that the proportion of interesting to mere "gap-filling" compounds is very small indeed ... If pentatriacontane, $C_{35}H_{72}$, were to disappear from print tomorrow it is safe to say that few would miss it, and no sensible chemist would suffer much grief over its loss.

These are not particularly cynical views on the matter of organic synthesis. They display a fairly common attitude between the two World Wars to the proliferation of new substances for the sake of it and a realization that such chemistry was in danger of becoming top-heavy with accumulated data and little supporting theory.

The insularity of German synthetic organic chemistry before the First World War was strengthened by the increasing specialism of the techniques involved. Organic chemistry has always shown this feature, notably in methods for isolation and purification (see TV programme 1), but it was augmented by the arrival of certain reagents which, though of great synthetic value, needed special kinds of handling. Two brief examples must suffice.

Ethyl acetoacetate, otherwise known as acetoacetic ester, had been obtained by the action of sodium on ethyl acetate and had been represented as a keto-form by Frankland and Duppa (1866) and as an enol by Geuther (1864). Modern theory regards it as a tautomeric mixture of both of them; their separation was accomplished in 1911 by Knorr:

$$CH_3 \cdot CO \cdot CH_2 \cdot COOEt \rightleftharpoons CH_3 \cdot C(OH) : CH \cdot COOEt$$
keto   enol

A detailed understanding of its chemistry dates from the work of Wislicenus in 1877. Apart from the interest as to its structure this compound featured in very many syntheses in the last century. Partly this sprang from the fact that the mobile hydrogen atom may be removed by strong bases (as OEt^- ions) to yield an anion that can then be alkylated (this, in modern terms, arises from the mesomeric stabilization made possible by the adjacent carbonyl groups):

$$CH_3 \cdot CO \cdot CH_2 \cdot COOEt \rightleftharpoons CH_3 \cdot C(OH):CH \cdot COOEt \quad \text{(tautomeric mixture)}$$

treatment with strong base, as OEt^-

$$CH_3 \cdot \overset{\|}{\underset{O}{C}} \cdot \bar{C}H \cdot \overset{\|}{\underset{O}{C}} \cdot OEt \longleftrightarrow CH_3 \cdot \underset{O^-}{C} : CH \cdot \overset{\|}{\underset{O}{C}} \cdot OEt \quad \text{(mesomeric anion)}$$

Alkylation takes place at the negatively polarized carbon thus,

$$CH_3 \cdot CO \cdot \overset{-}{C}H \cdot COOEt + CH_3I \longrightarrow CH_3 \cdot CO \cdot \underset{\underset{CH_3}{|}}{C}H \cdot COOEt + I^-$$

Hydrolysis of this product can then lead to either a ketone or an acid, depending on conditions:

$$CH_3 \cdot CO \cdot \underset{\underset{CH_3}{|}}{C}H \cdot COOEt$$

$$CH_3 \cdot CO \cdot CH_2CH_3 \qquad\qquad CH_3 \cdot CH_2 \cdot COOH$$

Alternatively the alkylated acetoacetic ester could then be alkylated a second time, and further ketones or acids obtained. A route of great generality to such compounds was thus opened up from acetoacetic ester.

The compound also had a multiplicity of functional groups which could be useful in effecting ring closure and other useful reactions.

> What reactive functional groups can you identify in the keto form of acetoacetic ester?

Carbonyl, $\overset{\diagdown}{\underset{\diagup}{C}} = O$

Carbethoxy, $-\underset{\underset{OEt}{|}}{C} = O$

Active methylene, $-CH_2-$, flanked by two electron-withdrawing groups.

One example of such syntheses is that by Knorr (1883) of 3-methyl-1-phenyl-pyrazol-5-one (used in the preparation of the drug antipyrine):

$$\begin{array}{c} CH_2 \!-\!\!-\! C \!-\! CH_3 \\ | \quad\quad \| \\ COOEt \; O \\ H \quad NH_2 \\ | \\ N \\ | \\ C_6H_5 \quad \text{phenylhydrazine} \end{array} \longrightarrow \begin{array}{c} CH_2 \!-\! C \!-\! CH_3 \\ | \quad\quad \| \\ CO \quad N \\ \diagdown \quad \diagup \\ N \\ | \\ C_6H_5 \end{array}$$

There were many more.

One other example of a synthetic reagent of wide applicability is the class known as Grignard reagents, discovered by Victor Grignard in 1900. These organomagnesium halides were again the subject of long structural arguments, but their utility as reagents was immediately recognized. Thus by 1905 there were more than two hundred papers dealing with them, and in 1912 (the year Grignard received the Nobel Prize) more than seven hundred. Despite being somewhat difficult to handle, Grignard reagents possess the enormous advantage of being able to attack a whole series of groups in organic compounds; they undergo replacement reactions with active hydrogen (OH, NH$_2$, etc.) and halogen, and addition to $\overset{\diagdown}{\underset{\diagup}{C}} = O$, $\overset{\diagdown}{\underset{\diagup}{C}} = N \overset{\diagdown}{}$ epoxy rings, etc. In sheer versatility they are probably still second to none (though in certain reactions alternative reagents, as organolithium compounds, are now more convenient).

It should not be supposed that organic chemists were so intoxicated with the joys of synthesis that considerations of utility played no part in their thinking: they certainly did, even within the context of purely academic research. As the 20th century approached there were signs that organic chemistry was reverting in some measure to its original quest, the study of natural products. Now, however, synthesis was to be pressed into service as one of the main approaches to the problem (the other being a study of the degradation products of the natural materials). The production of a 'natural' product from simpler materials

Figure 3.9 F. A. V. Grignard (1871–1935), French chemist famous for the discovery of the organomagnesium compounds named after him. While a research student under P. Barbier at Lyons he discovered that magnesium was readily attacked by alkyl halides in dry ether and that the product smoothly added to carbonyl groups to yield, ultimately, secondary or tertiary alcohols. The new Grignard reagents were to replace the zinc alkyls that had been, up till then, the most useful class of organometallic reagent. Within twelve years of their discovery (1900) more than 700 papers on their applications had been published. Grignard himself worked on little else, and most of his working life was spent where he first graduated, at the University of Lyons. In 1912 he received the Nobel Prize for chemistry.

of known constitution can often be decisive evidence for or against a proposed structure for it. The first major group of compounds to reveal its constitution was the carbohydrates, investigated by Emil Fischer in a series of papers from 1886 to about 1900. His systematic attack on the class was greatly facilitated by his use of the compound phenylhydrazine ($C_6H_5NHNH_2$) which he had discovered in 1875. This reacts with many sugars to form crystalline compounds known as osazones. These can be useful in identifying sugars but, since several sugars may form the same osazone, can also be used to relate their structures to each other.

This point may be briefly illustrated as follows for three of the hexoses (i.e. sugars containing 6 carbon atoms). Fischer wrote them as open chain polyhydroxy-aldehydes or ketones (modern theory regards them as usually existing as ring-structures in equilibrium with these). Osazone formation involved the first two carbon atoms of the chain:

$$
\begin{array}{ccc}
\text{CHO} & & \text{CH=NNHC}_6\text{H}_5 \\
| & & | \\
\text{H}-\text{C}-\text{OH} & \xrightarrow{3\text{C}_6\text{H}_5\text{NHNH}_2} & \text{C=NNHC}_6\text{H}_5 + \text{C}_6\text{H}_5\text{NH}_2 + \text{NH}_3 + 2\text{H}_2\text{O} \\
| & & | \\
\text{R} & & \text{R} \\
\text{sugar} & & \text{osazone}
\end{array}
$$

Now it so happened that three naturally occurring hexoses gave the same osazone: glucose and mannose (with an aldehyde group) and fructose (with a ketone group). Hence they must all have the same 'tails' (R in the above reaction).

Given the structure of glucose, that for mannose and fructose follows at once. It was this realization that gave Fischer the necessary impetus to tackle the appallingly complex task of elucidating the structures of the sugars. In his terminology, the reactions were:

$$
\begin{array}{ccc}
\text{CHO} & \text{CH}_2\text{OH} & \text{CHO} \\
| & | & | \\
\text{H}-\text{C}-\text{OH} & \text{CO} & \text{HO}-\text{C}-\text{H} \\
| & | & | \\
\text{HO}-\text{C}-\text{H} & \text{HO}-\text{C}-\text{H} & \text{HO}-\text{C}-\text{H} \\
| & | & | \\
\text{H}-\text{C}-\text{OH} & \text{H}-\text{C}-\text{OH} & \text{H}-\text{C}-\text{OH} \\
| & | & | \\
\text{H}-\text{C}-\text{OH} & \text{H}-\text{C}-\text{OH} & \text{H}-\text{C}-\text{OH} \\
| & | & | \\
\text{CH}_2\text{OH} & \text{CH}_2\text{OH} & \text{CH}_2\text{OH} \\
\text{glucose} & \text{fructose} & \text{mannose}
\end{array}
$$

phenylhydrazine *phenylhydrazine* *phenylhydrazine*

$$
\begin{array}{c}
\text{CH=N.NHC}_6\text{H}_5 \\
| \\
\text{C=N.NHC}_6\text{H}_5 \\
| \\
\text{HO}-\text{C}-\text{H} \\
| \\
\text{H}-\text{C}-\text{OH} \\
| \\
\text{H}-\text{C}-\text{OH} \\
| \\
\text{CH}_2\text{OH} \\
\text{osazone}
\end{array}
$$

(These are Fischer's projection formulae which, by a convention we do not need to describe here, define uniquely the configuration of each asymmetric carbon atom. The three sugars are today given the prefix D- to denote the configuration of the lowest asymmetric centre.)

It has been said that Fischer's work in carbohydrate chemistry constituted the first great test and vindication of the hypothesis of the tetrahedral carbon atom. Without that hypothesis his work made no sense at all.

When this work was nearing completion Fischer turned to the more complex substances known as proteins, showing how individual aminoacid residues may be linked together into a chain to form a peptide—a kind of very simple protein.

Figure 3.10 Emil Fischer (1852–1919), German organic chemist and pioneer of natural product chemistry. A student of Kekulé and Baeyer, he occupied chairs in several German universities. Using phenylhydrazine, which he had discovered (1875), he unravelled many of the problems of carbohydrate chemistry and was eventually able to define the configuration of all 16 aldohexoses, $CH_2OH.(CHOH)_4.CHO$. Another major triumph lay in the field of protein and peptide structure. He showed how to tackle the problem of amino-acid sequences and was able to link no less than 18-amino-acid residues in one synthetic peptide. Other work included syntheses of heterocyclic compounds and examination of the naturally occurring group which he termed 'purines', based upon uric acid.

In 1907 he synthesized a peptide with 18 amino-acid residues, at the time a most remarkable achievement. Other groups to be intensively examined in the early years of this century were terpenes (including camphor, which was synthesized by Komppa in 1903), alkaloids, plant pigments (including the anthocyanins and chlorophyll), rubber and one or two other major classes. To all of the investigations the words of Fischer in 1907 are applicable:

> Laboratory synthetic methods will be indispensable for a long time to come, not only for preparative purposes but also as a means of elucidating the structure of complex substances of natural origin.

SAQ 10 (*Objective 6*) What role did these syntheses play in the total structure of chemistry? Was it convergent or divergent?

3.1.3.2 Theoretical organic chemistry

By 1900, organic chemists could look back on 30 or 40 years of spectacular synthetic work, in the course of which many anomalies had been resolved and the suppositions of the structure theory abundantly justified. Yet, in fact, there had been little advance in theory commensurate with the body of facts assembled. In terms of its ethos and ideology, organic chemistry must have ranked as one of the most conservative branches of science at that time. Yet there *were* many things still to be explained. Most obvious, perhaps, were the problems associated with benzene. The non-existence of more than three disubstituted benzenes was still raising doubts about Kekulé's simple hexagon, for according to that *o*-dichlorobenzene (for example) should have two forms and only one was ever found. Kekulé's oscillation hypothesis was regarded by many as contrary to the whole spirit of structure theory in making two formulae represent one substance. According to H. E. Armstrong 'this oscillation hypothesis never found favour':

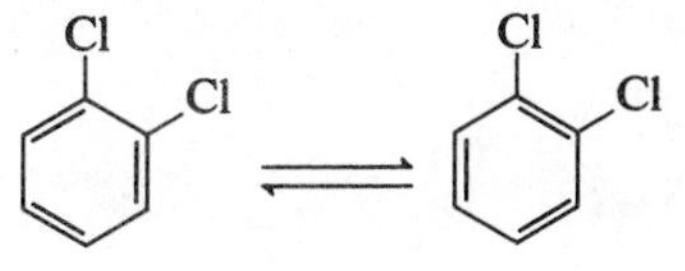

The number of alternative suggestions was vast. No one at that stage realized that the benzene formulae littering the journals in the late 19th century were attempts to express what was, in classical terms, inexpressible. The most successful—because the least constrained by the accepted theories of valency—was that of J. Thiele in 1899. Indebted, like Werner, to earlier ideas of 'residual affinities' he used them to explain the reactivity of the organic double bond. The adding reagent would find an anchorage there because of the pre-existence of 'partial valencies' left over when the double bond was formed:

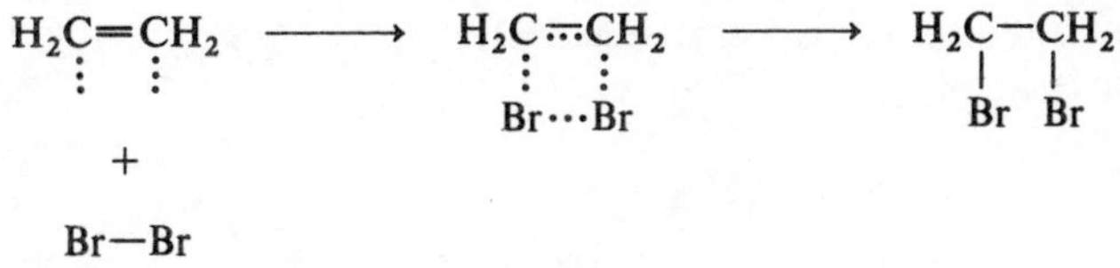

In this way he explained the phenomenon of 1:4 addition to butadiene, the central pair of partial valencies effectively neutralizing each other:

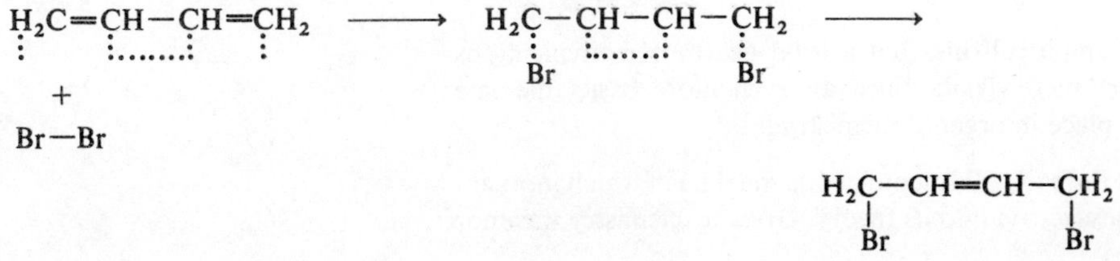

Application of this to benzene not only gives a hexagon of considerable symmetry but also goes some way to explaining benzene's lack of reactivity towards addition reagents:

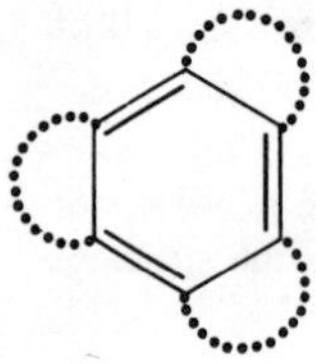

However, by the same token, one would expect low olefinic reactivity in both cyclooctatetrene and cyclobutadiene. The first was shown (by Willstätter in 1911) to be intensely reactive and quite unlike benzene; the second resisted all attempts at capture until 1965, and long before that was realized to be even more reactive.

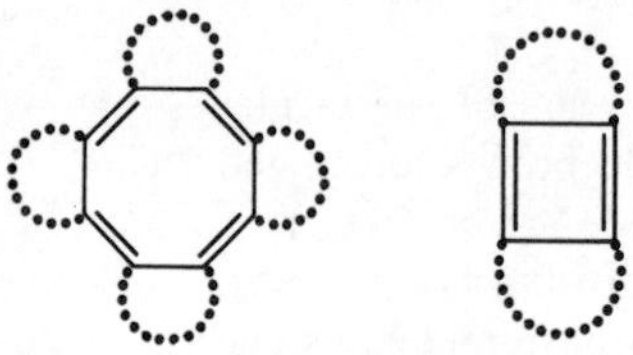

Hence Thiele's theory, at least on its own, could not be an adequate explanation for the uniqueness of benzene.

Other areas of real difficulty included the apparently outrageous case of trivalent carbon in the triphenylmethyl radical, discovered by Gomberg (in 1900) from what was then thought to be the reversible dissociation of hexaphenylethane (since 1968 it has been known that the radical dimerizes to a different product):

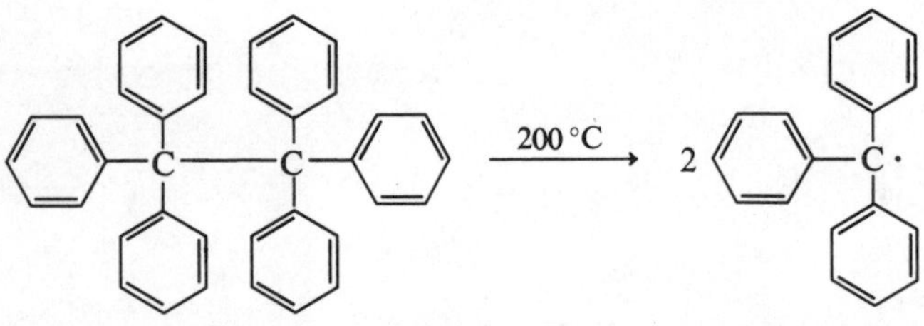

Ironically this discovery, so embarrassing to established valency theory, was the first genuine case of a free radical to be isolated, the culmination of the dreams of Berzelius, Frankland and many others whose work lay at the foundation of that theory.

Problems of structure were fundamental, but problems of reactivity were hardly less so. Again, benzene chemistry offers the most pressing examples. Why do certain groups already present in the aromatic molecule seem to direct further incoming groups into specific positions? Why do some groups, as OH, NH_2, CH_3, Cl, etc., direct to the *ortho* and *para* positions, whereas others, as NO_2, COOH, CN, etc., direct to the *meta* position?

For example, on nitration,

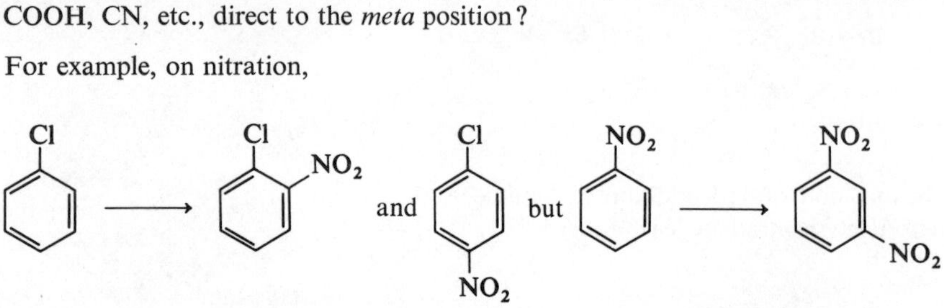

There was no shortage of empirical rules but a total dearth of convincing explanation. These difficulties merely underlined an even more basic dilemma: why did *any* reaction take place in organic chemistry?

Now it will be recalled that the early 20th century was the time in which inorganic chemistry was accepting the electron into its theory. Organic chemistry was more

24

reluctant. Again we can learn from that invaluable commentator on the chemical scene, A. W. Stewart (1918):

> Though several attempts have been made in this region of the subject, organic chemists have not welcomed them with anything like wholehearted encouragement. There is a feeling, apparently, that in abandoning the usual structural formulae and replacing them by electronic symbols the subject is being complicated rather than simplified; and this feeling, whether it is due to scientific caution or to mere conservatism, has certainly carried the day for the present.

Some of the earliest of these attempts had run into exactly the same kind of trouble Berzelius had encountered when he tried to relate ideas derived from electrolytes to substances that were essentially non-ionic. The American chemist H. S. Fry in 1908–9 proposed a dualistic view of what we should call covalent bonds, regarding the participant atoms as having acquired a complete charge. Thus benzene was written as

and this does emphasize the similarity between *ortho* and *para* positions. But such extreme views never found general favour, mainly, it would seem, because they glossed over the essential differences between ionic and non-ionic compounds and appeared to have no basis in physical data.

Meanwhile an alternative approach was being developed by the English chemist A. Lapworth, originating in pre-electronic days. In the first few years of this century he developed two extremely important ideas. In studying the action of the weak prussic acid on camphorquinone he noted the rate of reaction by observing the rate at which the yellow colour of the quinone disappeared:

$$\text{camphorquinone} + \text{HCN} \longrightarrow \text{cyanhydrin}$$
$$\text{(yellow)} \qquad\qquad\qquad\qquad \text{(colourless)}$$

He found that acids and alkalis had a striking effect on the rate. Whereas a pure aqueous acid required many hours for the reaction to be completed, the addition of hydrochloric acid seemed to cause it to be postponed indefinitely, while the addition of a little alkali gave instantaneous decolorization.

What do these results suggest about the nature of the reagent?

> Since HCN is a weak acid the cyanide ion concentration will be extremely small; addition of a strong acid will decrease it still further, because this increases the concentration of hydrogen ions. But an alkali will promote formation of cyanide ions by the reaction
>
> $$\text{HCN} + \text{OH}^- \longrightarrow \text{H}_2\text{O} + \text{CN}^-$$
>
> Consequently the cyanide ion must be the reagent in this case.

From this work Lapworth began to develop a generalized theory of active reagents, recognizing that the active species may not at all be the same thing as the name on the reagent bottle. In this case he assumed the carbonyl group must in some way be polar, with the negatively charged cyanide ion attacking the positively polarized carbon, thus:

Lapworth also noted that alternate atoms in a conjugated chain often show considerable similarity in chemical behaviour and proposed his α-γ rule:

> The alternate atoms in such a chain might be expected to exhibit similar powers of acting as seats of ionic activity.

Figure 3.11 Arthur Lapworth (1872–1941), English chemist noted for pioneer studies in the mechanism of organic reactions. After various posts in London, he moved to Manchester in 1909 as Senior Lecturer in inorganic and physical chemistry, later succeeding Perkin as Professor of Organic Chemistry. His expertise in all three branches showed itself in his mechanistic interpretation of reactions in the camphor series. His classification of reagents (corresponding to Ingold's later *electrophilic* and *nucleophilic*) laid the foundation of modern electronic theory of organic chemistry.

This effect was later ascribed to a 'key atom' which was the source of such electronic disturbances. Thus, taking the nitrogen as the key atom in the bromomethylbenzyl cyanides he showed how the sluggish behaviour of the bromine in the *meta* isomer could be explained. It is likely to be removed as Br^-:

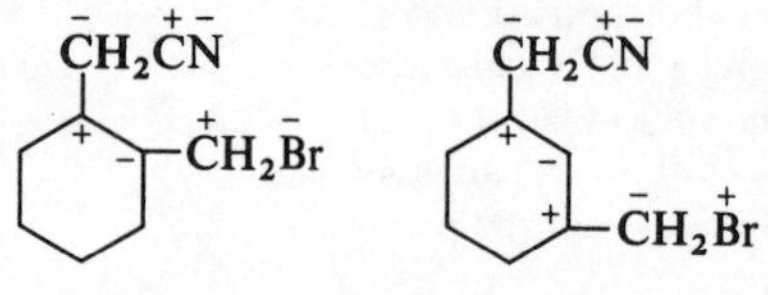

$$\overset{-}{C}H_2\overset{+}{C}\overset{-}{N} \qquad \overset{-}{C}H_2\overset{+}{C}\overset{-}{N}$$

By 1920 Lapworth presented to the Manchester Literary and Philosophical Society an important paper in which these ideas were explained in terms of the electronic theory of valency. Unlike Fry, however, he pointed out that the + and − signs are not regarded as representing fully electronic charges but 'merely as expressing the relative polar characters which the two atoms seem to display at the instant of the chemical change in question'. Thus he wrote

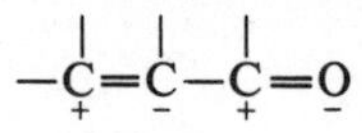

$$-\overset{+}{\underset{}{C}}=\overset{-}{\underset{}{C}}-\overset{+}{\underset{}{C}}=\overset{-}{O}$$

and

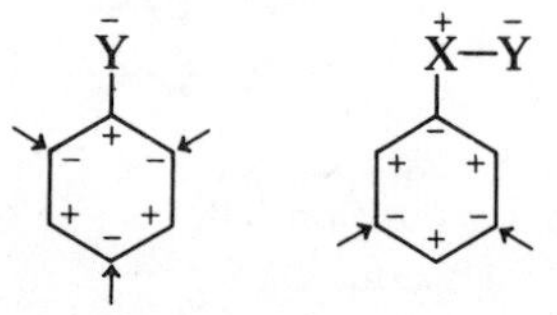

These views were taken further two years later by Kermack and Robinson. Instead of postulating an alternating series of polarities, however, they used curved arrows to represent displacements of electron pairs

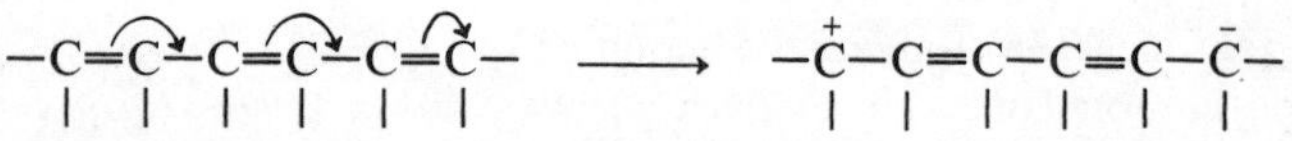

$$-C=C-C=C-C=C- \longrightarrow -\overset{+}{C}-C=C-C=C-\overset{-}{C}-$$

At first these ideas were met with some scepticism; H. E. Armstrong observed 'A bent arrow never hit anything'. The state of confusion was acknowledged by Robinson when, as President, he introduced a Faraday Society discussion in 1932 on the application of electronic theory to organic chemistry. He said:

> At present it would seem as if we are in a transition stage of knowledge comparable with that which obtained in the 'forties and 'fifties of last century, and the chemists of two or three generations hence will look back upon the present confusion with the same feelings as we experience in regarding that time.

But two events were rapidly to change the situation. One was the general acceptance of the quantum theory. This was later to provide a detailed rationale for much of the empirical rules of the new electronic theory, and even in the late 1920s enabled N. V. Sidgwick to produce his influential and confident book *The Electronic Theory of Valency*. Secondly there was a rising awareness among organic chemists that the classical structure theory was increasingly unable to account for many of the new phenomena being reported. Thus, in 1924, F. G. Arndt insisted on the representation of γ-pyrones by two formulae simultaneously—a revolutionary suggestion that found its full fruition in the later theory of resonance:

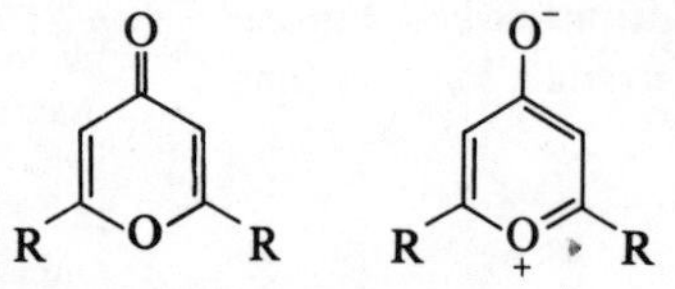

In the same year C. K. Ingold began a six-year programme of research at Leeds, in which he developed the ideas of Robinson and Lapworth and, among many other things, distinguished between permanent and temporary elements in polarization of covalent bonds. From this work, and from that of several other research centres in Britain and the United States, a new theory of organic reactivity emerged in which concepts such as the inductive effect, resonance (mesomerism), etc., entered the vocabulary and thinking of most British organic chemists.

26

It is not possible to offer more than the briefest reference to these events, but it should be clear that they were giving back to organic chemistry the coherence and confidence that had marked its high progress in the 19th century and which had been temporarily lost in the years before the First World War. The debt to inorganic chemistry was small, but a state of increasing dependency upon the physical chemists was being manifest. It was this that was to usher in the new convergent phase with which we conclude this Unit.

> **SAQ 11** (*Objective 6 and revision*) The following statement could have been made in 1831, 1871, 1901 or 1931:
>
>> On the theoretical side organic chemistry seems at first sight to have reached an almost unnatural perfection so far as main principles go, for few sciences have attained such a high pitch of organisation in so short a period.
>
> In which of these years do you think the quotation originated?

3.1.4 The emergence of physical chemistry

A fourth feature of chemistry from about 1875 to the 1940s was the emergence of a new science: physical chemistry. We do not intend to discuss this in any detail here, but must note briefly how it arose as a separate specialism, how it fared in relation to the two well-established branches of chemistry, and how it affected the relations between those branches.

3.1.4.1 The birth of a third chemical specialism

It is obvious from a consideration of the 'indicators' mentioned earlier that in the 1880s the science of physical chemistry had achieved sufficient recognition to be awarded comparable status to the older two branches of organic and inorganic chemistry. This was to be an amalgam of thermodynamics, reaction kinetics, electrochemistry, surface chemistry, spectroscopy and other studies of essentially physical properties of matter. Such studies did not, of course, originate at this time. Even the term 'physical chemistry' dates back to the 18th century, a period when chemists were inspired by a vision of a quantified science comparable to physics in its definiteness and precision. For chemistry, however, no Newton was to arise then or even later, and the culmination of the 18th century efforts in this direction came in 1803 with the *Essai de Statique Chimique* of C. L. Berthollet. The author (not to be confused with Berthelot) emphasized the importance of mass in chemical reaction, suggesting that forces of affinity, like those of gravitation, are proportional to the masses of the substances reacting. His position was later to be vindicated in the general acceptance of the law of mass action.

The 19th century saw the growth of knowledge in many areas that we should today call 'physical chemistry', but a point that is liable to escape the modern observer is that the gap between chemistry and physics was far less at the beginning of that century than it was at the end. Indeed it is inappropriate to designate as either 'chemist' or 'physicist' many of those who contributed to both sciences. Dalton, Davy and especially Faraday defy attempts to categorize them in this way. So, to a lesser extent, do men like Bunsen and van't Hoff. They were able to use their physics to inform their chemistry, and vice versa, because they had a foot in each camp. By the end of the century, however, the gap had widened so far as to make such intellectual gymnastics much more difficult. It has often been said that by then physicists and chemists scarcely spoke to one another, while both would consider biologists as hardly scientists at all. The sheer volume of accumulated knowledge in one science was an effective deterrent against dabbling in another, and increasing isolation was inevitable. Only the strongest constraints could overcome such centrifugal tendencies in science. For physics and chemistry these were, in due course, to come from the discoveries of radioactivity and the nuclear atom, but that was only after about 1900. Meanwhile, those chemists who were aware of the insights available to chemistry from physics were, perhaps unwittingly, regrouping to form the nucleus of the new 'physical chemistry'.

Physical chemistry as a specialism may be considered the inevitable product of the expansion and divergence of the two parent sciences, much as biochemistry has become in our own day. Its separate existence is owed to van't Hoff, Arrhenius and (above all) Ostwald. At Leipzig, Wilhelm Ostwald was the archetypal physical chemist, concerned not with structures but with energetics (he even called his country home *Energie*), proud to have discovered no new compounds whatever, and dedicated to the elimination from chemistry of crude mechanical models of atoms and molecules. In Ostwald's hands the subject became rapidly professionalized. The thermodynamics that he derived in part from the American J. W. Gibbs required a different kind of expertise and attitude from those traditionally associated with chemists from the other two branches and the new kind of chemist had to be specially bred. Ostwald's research school became world-famous and received many visitors from overseas, particularly from the United States, to which, in due time, Ostwald's methods were transplanted. His own *Zeitschrift für physikalische Chemie* (1887) and his two great volumes *Lehrbuch der allgemeinen Chemie* helped to propagate his views and establish the new subject. We must now enquire what were the consequences for the practice of chemistry as a whole.

3.1.4.2 Consolidation of the new discipline

It is possible to imagine a situation in which the new physical chemistry might have developed for a while and then been absorbed within either physics or the two older branches of chemistry. In fact, nothing like that took place and, for the next few decades, the physical chemists were working in some isolation from their other colleagues. Several reasons suggest themselves for this.

Figure 3.13 Ostwald (right) and van't Hoff (left; see Unit 2, Figure 2.25), co-founders of physical chemistry. F. W. Ostwald (1853–1932) was a German chemist who from 1887 occupied the Chair of Physical Chemistry at Leipzig. Accepting the ionic dissociation theory of Arrhenius he applied mass action considerations to arrive at his famous dilution law. Using simple apparatus (as shown here) he did extensive work on the physical chemistry of solutions. He wrote much, favouring a thermodynamic rather than an atomistic approach to chemistry. A co-founder of *Zeitschrift für physikalische Chemie*, he also initiated his series of classics in the history of science, Ostwald's *Klassiker*. He received the Nobel Prize for chemistry in 1909.

Figure 3.12 J. W. Gibbs (1839–1903), American chemist and one of the founders of chemical thermodynamics. Gibbs began academic life as a mathematician and engineer, and in 1871 obtained an unsalaried Chair of Mathematical Physics at Yale. Having sufficient private income of his own, he was able to retain this post till his death. His first paper on chemical thermodynamics (1876) dealt with the phase rule and was brought to the attention of chemists by Clerk Maxwell. He also published on statistical mechanics and thermodynamics of fluids.

It is quite clear from the chemical literature that a considerable body of scepticism existed towards physical chemistry until well into the present century. Once again we encounter the problem of *analogy*. Among others, in the 1880s van't Hoff drew attention to the analogy existing between the osmotic pressure of a dilute solution and that of an ideal gas (i.e. one that obeys the general gas law $PV = RT$ completely). Denying that this was 'a fanciful analogy' he claimed that the osmotic pressure of such a solution would be equal to the pressure the solute would exert if it alone existed as a gas having the same volume as the solution. This was simply too much for chemists accustomed to thinking in literal terms of the kinetic behaviour of gases and tended to be rejected out of hand. Then there was also the problem of *anomaly*. The fact was that vast numbers of cases were reported in which some of the laws of the new physical chemistry were seen to break down. We can only mention one, the famous (or infamous) case of Ostwald's dilution law. This was an application (1888) of the law of mass action (Guldberg and Waage, 1867) to the theory of electrolytic dissociation of Arrhenius (1883). Arrhenius had proposed that solutions of electrolytes contained the solute in at least a partly ionized form. The degree to which ionization had occurred could be deduced from conductivities of such solutions. Ostwald's reasoning was on the following lines.

Let one mole of an electrolyte AB be dissolved in a volume of V litres, and suppose that α moles dissociate into ions; $\alpha(<1)$ is the degree of dissociation:

$$\text{A--B} \rightleftharpoons \text{A}^+ + \text{B}^-$$

There will then be these concentrations:

$$\text{A}^+, \alpha/V; \quad \text{B}^-, \alpha/V; \quad \text{AB}, (1-\alpha)/V$$

Since by the law of mass action

$$\frac{(\text{concentration of A}^+) \times (\text{concentration of B}^-)}{(\text{concentration of AB})} = K$$

where K is the ionization constant,

$$\frac{\dfrac{\alpha}{V} \cdot \dfrac{\alpha}{V}}{\dfrac{1-\alpha}{V}} = K$$

hence

$$\frac{\alpha^2}{(1-\alpha)V} = K$$

which is the dilution law. If α is very small compared with unity this approximates to

$$\frac{\alpha^2}{V} = K$$

or

$$\alpha = \sqrt{KV}$$

and so the degree of ionization will vary with the dilution, and a graph of α against $\sqrt{V}$ (or $1/\sqrt{\text{concentration}}$) should be a straight line.

When Ostwald published his law he illustrated it with experimental data for five organic acids, all of which are weak and therefore have low values for α. However, within a year it was widely realized that the dilution law was not applicable to strongly ionized electrolytes, even without using the approximation in the derivation above. Thirty-five years were to elapse before the anomalous behaviour of strong electrolytes was to be explained by Debye and Hückel in 1923 (and then only for very dilute solutions). As J. H. Wolfenden has recently commented, this 'was a problem which only physicists could solve and in whose solution only chemists were interested'. Yet the absence of a satisfactory solution from the physical chemists did not endear their approach to many of their inorganic and organic counterparts. After all, it was a fairly fundamental deficiency since so much of physical chemistry was based upon the Arrhenius/Ostwald views of electrolytic dissociation. When, in 1896, G. F. Fitzgerald gave a Memorial Lecture for the lately deceased Helmholtz he went out of his way to deliver a broadside on much of the physical chemistry of the day, including the theory of solution, electrolytic dissociation, osmosis and the application of thermodynamics to chemical investigations where there are 'serious pitfalls into which investigators have fallen'.

Figure 3.14 S. A. Arrhenius (1859–1927), Swedish chemist, famous with Ostwald and van't Hoff for the foundation of physical chemistry as a reputable discipline and also for his own theory of ionic dissociation. This theory appeared in embryo form in his doctoral thesis at Stockholm (1883), but it was poorly received and it took nearly five years of travel, correspondence and lobbying for Arrhenius to get it generally accepted. This done, he became a major international figure in physical chemistry, and in 1903 he was awarded the Nobel Prize for chemistry.

It must not be assumed that scepticism was all on one side. Carried away with enthusiasm for the new physical chemistry which came to definite conclusions independently of any particular models of reality, Ostwald sought to purge away all mechanical hypotheses from chemistry. Adopting a positivist philosophy which bore a superficial resemblance to that of Berthelot, he became a bitter opponent of atomism, regarding with profound scepticism the efforts of organic chemists, particularly, to understand their phenomena in atomic terms. Writing in 1904—more than a hundred years since the first appearance of Dalton's atomic theory—he could assert:

> Chemical dynamics ... has made the atomic hypothesis unnecessary for this purpose and has put the theory of stoichiometrical laws on more secure ground than that furnished by a mere hypothesis.

Such two-way scepticism, going well into our century, inevitably hardened the barriers between physical chemistry and the rest of the subject. But perhaps the most important contributor to this process was the sheer inability of organic and inorganic chemists to see any relevance in what the physical chemists were doing to the problems *they* were trying to tackle. This was no new thing. Back in the mid-19th century Kopp and others had sought for connections between physical constants of substances and their constitution, but their success had only been indifferent, partly because in many cases no simple relation exists. Until the 1930s the reluctance of organic and (to a smaller degree) inorganic chemists to draw freely upon the resources of physical chemistry is conspicuously evident. Of course there are exceptions, including the American chemist A. Michael, who, around 1900, made strenuous efforts to interpret organic reactions thermodynamically. Yet the contribution for which he was and is best known is the synthetic reaction named after him (involving the addition of the sodium derivative of acetoacetic ester or a related compound to α,β-unsaturated esters or ketones). But Michael was unusual. L. H. Sutton has related how organic chemists in Britain were unable to use physical techniques to help them solve problems of structure:

> In general, in the early 1920s there was little that one could do by physical means to ascertain the structure of a compound save to measure its molecular weight and speculate.

He also records W. H. Perkin's not untypical attitude to the attempts by Sidgwick to go somewhat further than this:

> Perkin regarded Sidgwick's activities as trivial; he is said to have remarked that 'physical chemistry is all very well but it does not apply to organic substances', although Sidgwick may have invented this, or at least improved on the original.

We shall see in Section 3.2 how eventually physical chemistry *was* found to 'apply to organic substances' but until the Second World War the issue was in some doubt. Meanwhile, another consequence of the rise of physical chemistry requires a short comment.

3.1.4.3   Impact of physical chemistry on the organic/inorganic relationship

This may be put very simply in the following way. In so far as physical chemistry was able in its first 60 years or so to take care of topics traditionally in the province of either of the other two branches, it did so by syphoning off precisely those parts of the subjects that might have provided a common link between them. Both organic and inorganic chemists were facing a number of common problems, but a united assault was rendered less likely by the tacitly agreed removal of such matters to the domain of the physical chemist.

The study of weak electrolytes in connection with Ostwald's dilution law was not confined to organic acids; inorganic compounds furnished other examples. Research on reaction rates covered hydrolysis of esters, the peroxide oxidation of hydrogen iodide, and many other examples drawn from both branches. Photochemical work (involving the action of light on chemical reactions) ranged from the deceptively 'simple' combination of hydrogen and chlorine to the obviously complex process of organic photosynthesis. Systems containing metals, sulphur and sulphides were as much the objects of attention by the colloid chemists as were those derived from starch, soap and gelatin. Adsorption on charcoal was as important for simple inorganic gases as for complex organic dyestuffs, and both were studied. Examination of catalysis (for which Ostwald received a Nobel Prize) was not limited to the numerous cases of metallic oxides promoting the decomposition of potassium chlorate; enzyme-catalysed hydrolyses of carbohydrates were just as appropriate.

And so one could go on. The point is that physical chemists, with their newly developed mathematical and experimental skills, were far more likely to be found engaging in this kind of research than those whose expertise lay elsewhere, in preparing and purifying individual compounds. In this way physical chemistry became a kind of wedge inserted between the other two areas and beyond doubt helped to keep them apart for many more years.

> **SAQ 12** (*Objective 7*) One of the important advances in thermodynamics came in 1906 with the announcement of the so-called 'Nernst heat theorem' by the Professor of Physical Chemistry in Berlin, H. W. Nernst. From this came the third law of thermodynamics which may be expressed in several ways, e.g.: *In a system where all factors are in internal thermodynamic equilibrium, their entropy will become zero at absolute zero.* Thus at that temperature a perfectly crystalline solid would have zero entropy. Details may be found in the usual textbooks if you are interested, but they are not necessary for the present SAQ. Nernst's paper concluded with these words:
>
> > If we summarise the results of our considerations briefly, we can say that the final goal of thermochemistry, namely the exact calculation of chemical equilibria from heat effects, seems possible if we take the new hypothesis as an aid, according to which the curves of free energy and total energy of chemical reactions between pure and liquid bodies meet at absolute zero. For the time being there is not enough information for the

Figure 3.15 W. H. Perkin (1860–1929), English organic chemist. The son of the elder W. H. Perkin (see Unit 2, Figure 2.4), he taught successively at Edinburgh (Heriot-Watt College), Manchester and Oxford. He made important contributions to the study of natural products, including terpenes, alkaloids, and dyestuffs. He was the first to make a 4-membered ring, in contradiction of the strain theory of his former teacher Baeyer.

formulae derived by the aid of this hypothesis to be exactly tested. Data for specific heats at low temperature are not available, and so in this work we have derived approximate formulae that can be tested in practice under varied conditions. The connexion between heat and chemical affinity appears to be essentially clear and is obviously important.

What kind of impact do you think this paper would have had on organic and inorganic chemists of the day?

3.1.5 External factors

So far we have been considering the growth of a specialist chemistry, with inorganic, organic and later physical chemistry tending to pursue independent and even diverging courses. The reasons, we have suggested, may be found within the confines of chemistry itself. In this final Section devoted to the phase of *specialization* we glance briefly at some of the events in the world outside that impressed their mark upon chemistry. These are the growth of industry, the varying patterns of education, and some subtle changes in ideology.

3.1.5.1 Industry

The meteoric rise of the organic chemical industry is one of the formative events in modern European history. It is, of course, in large measure the story of dye-stuffs manufacture in Germany. By 1900 the great German firms of Hoechst, BASF and others were accounting for 90 per cent of world dyestuff production, and this does not include German-owned plants in France, Switzerland and elsewhere. As the Harvard economic historian D. S. Landes has written:

> In technical virtuosity and aggressive enterprise, this leap to hegemony, almost to monopoly, has no parallel. It was Imperial Germany's greatest industrial achievement.

The economic and military advantages conferred by the possession of this industry played a decisive role in the early part of the First World War.

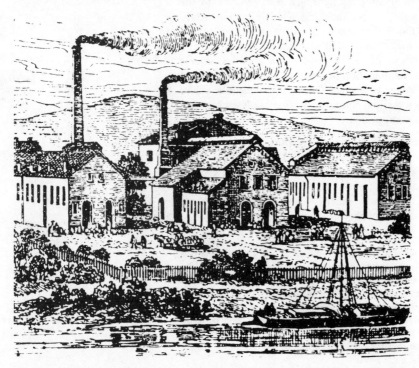

Figure 3.16 The Hoechst factory at Frankfurt-am-Main in 1865. This marked the beginning two years earlier, of one of the great German chemical industrial enterprises. The product ('Fuchsine') was a red dye obtained from the oxidation of *very* impure aniline, and was a derivative of triphenylmethane. The equipment was a small boiler and a 3 hp steam engine, with a labour force of five workmen, one accountant and one chemist. The plant was demolished in 1874; its successor now employs more than 50 000 people.

Thus we are confronted with yet another manifestation of the importance of organic synthesis. Although the industrial exploitation of this may be said to have begun in England (with Perkin's production of mauve in 1856), the lead rapidly passed to Germany for precisely the reason that here organic chemistry was a specialism based upon a strong research foundation. Whereas the Badische Anilin und Soda Fabrik at Ludwigshafen (BASF) employed five research chemists in 1870, by 1884 they had sixty-one. No British or French firm could possibly compete at that level. When, in 1869, synthetic alizarin was simultaneously produced by Perkin in England and by Graebe and Liebermann in Germany, its manufacture was started by both BASF and Hoechst, while Perkin had to make his under licence from BASF, and the complete anthracene production of one English company of coal-tar distillers was sold to Hoechst for conversion to alizarin! Many years later, in 1914, the British troops went to war in uniforms whose khaki dyes were imported from Germany but derived from coal-tar chemicals originating in Britain.

Partly responsible for the early proliferation of synthetic methods was the chaotic situation with regard to patent rights. Many smaller chemical firms were being destroyed through engaging in costly litigation over patents, and there was salutary example of the great French dyestuffs firm La Fuschine, which on this account went into bankruptcy in 1868. The only safe policy was to outdo one's competitors in new methods and products on as large a scale as possible, so massive research programmes were launched, resulting in literally thousands of new dyestuffs. A classic case was that of indigo, for which a whole succession of different synthetic routes was developed in the 1880s, partly to keep the price as low as possible but partly to avoid patent infringements.

Almost as a by-product of dyestuffs another development occurred in the late 19th century in which the German chemical industry began to produce a steady output of pharmaceuticals. Following several antipyretics (drugs that will reduce abnormally high body temperatures) aspirin was introduced in 1899. The science of chemotherapy owed much to the efforts of the German physiologist Paul Ehrlich, who set out to test many new compounds for possible therapeutic action. His search for an antisyphilis drug culminated in the discovery of the organo-arsenic compound salvarsan, though only after 605 other materials had been tried and found wanting. So, for very different reasons, drugs joined dyes as industrial products calling for extensive and protracted research in the synthesis of organic compounds. While this remained true the existence of organic chemistry as a defined specialism could hardly have been in doubt.

After the First World War the situation changed, in that Germany was now deprived of patents, trade-marks, trading opportunities and other incentives to continue as before. Organic syntheses were taken up again, but now they were accompanied by research in quite different areas. The shortage of coal-tar led to a new appreciation of the possibilities of petroleum as a source for organic compounds and the emphasis began to shift from aromatic to aliphatic chemistry. In inorganic chemistry vast numbers of new compounds were not required, but new methods of making the dozen or so important 'heavy chemicals' were necessary in Germany and elsewhere in a postwar Europe, where inflation was necessitating economies of a severe nature. Karl Winnaker, chief executive of Hoechst for sixteen years, recaptured the spirit of those days:

> Physical chemistry and electrochemistry, which had such modest beginnings, recorded great success only in the last years before the [1914] war when they excited tremendous interest and were greatly developed. Catalysis and high pressure techniques were milestones on the road of this development. All this epoch-making progress had now to be adapted for a peacetime economy. Anyone in industrial or university research looking for a new career was forced completely to re-orientate his previous work—economically, methodologically, and even scientifically. This might have been tremendously attractive—as in fact it was—but it took place against the backdrop of a hopeless political and economic situation.
>
> (K. Winnaker (1972) *Challenging Years*, trans. D. Goodman, Sidgwick and Jackson, London)

The outcome tended not to be research in what was recognizable as inorganic chemistry so much as in the comparatively new physical chemistry. Research into catalysis leading to the contact process for sulphuric acid had marked the entry of BASF into physical chemistry as far back as 1890. Just before the First World War the manufacture of synthetic ammonia by the catalytic combination of hydrogen and nitrogen at very high pressures had led Haber and Bosch the same way. Now, with the war over, these lines of approach were investigated with redoubled energy. Nor did they lead just to new methods for inorganic chemicals. Between the wars, German chemistry held the lead in the application of high pressures to many gaseous reactions involving organic compounds, leading to (among many other things) synthetic alcohols from carbon monoxide and hydrogen, synthetic petrol from the same reagents but under different conditions, and a wide range of products obtainable from acetylene. Thus the outcome was a new concentration on that branch of chemistry dealing with the general principles of reactions, especially under unusual conditions—and that, of course, was physical chemistry. One must see this as a strong impetus to the growth of that branch of the subject. But there was still specialization. It is amusing to recall that, when Ostwald was invited to address BASF in 1928, he shocked the management by choosing as his subject: 'The organization of progress—or how to render the specialist harmless'. That at least was how he saw the shape of industrial chemistry in his time.

3.1.5.2 Education

The increasing specialism of the German chemical industry from the 1870s could only have been possible on the basis of a highly specialized training programme. A shrewd observer of the German chemical scene over a very long period was A. W. Stewart. The comment below was made in 1931, but is consistent with opinions he had expressed well before the First World War:

> Since the time of Kekulé, organic chemistry has been for the most part a synthetic science. At the present day considerably over a hundred thousand organic compounds are known, and one need not have the least hesitation in saying that if seventy per cent of them had never been synthesized we should not feel the lack of them to any appreciable extent. The origin of this enormous flood of synthetic material is to be found in the German University system; for since, under the German regulations the degree in chemistry is granted only on the results of original research, it follows that every Ph.D. in organic chemistry represents so many new compounds—at least as a general rule.

Thus the specialization in organic chemistry shown by German industry may be seen as a reflection of the same tendency in the educational system. And of course industry encouraged it. It had every reason to do so, as the graphs (Figures 3.17–3.19) will demonstrate.

> **SAQ 13** (*Objective 8*) (a) Where does industrial production seem largely independent of educational advance? Why?
>
> (b) Which of these two items is more sensitive to changing political situations? Why?
>
> (c) Could you make any future predictions on the basis of these graphs?

Inevitably such comparative data prompt an enquiry into the state of British scientific education from around 1870. It is obvious from the graph (Figure 3.17) that its growth-rate was always much smaller than that of Germany until the postwar depression of the 1920s. However, we are more concerned here with the nature of that education rather than its extent—almost with quality rather than quantity. And in that connection we are faced with a most intriguing problem of specialization.

At about the middle of the last century demands were being made for a much more specialist type of scientific education. It was reported that Tyndall wanted a degree in 'heat'! But he was fairly exceptional, and most forward-thinking educators were arguing the other way. Lamenting the traditional English public school preoccupation with classical studies they urged that education should be much broader, and should include the sciences. The notion of a comprehensive and liberal education was urged even by the Devonshire Commission set up in 1870 as a Royal Commission on Scientific Instruction and the Advancement of Science. This body recommended that, in the universities, science should not be specialized, advocating the study of other subjects as well. This was in the

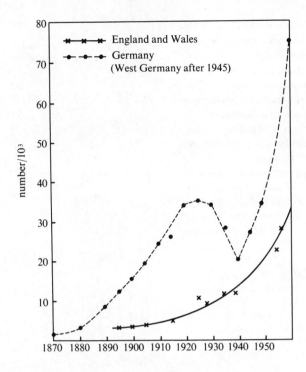

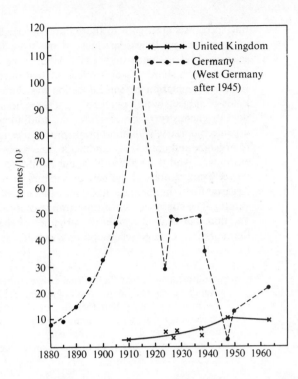

Figure 3.17 (above) Students of science and technology at University in (a) England and Wales, and (b) Germany, 1870–1965. This includes British University Colleges and German Technische Hochschülen.

Figure 3.18 (above right) British and German exports of dye-stuffs, 1880–1965.

Figure 3.19 (right) British and German production of sulphuric acid, 1870–1965.

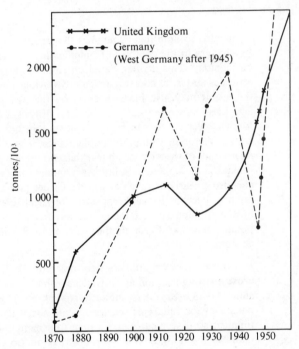

growing tradition that included many chemists of great distinction who deplored the 'one-sidedness' of many scientists (to use Hofmann's word), and who advocated precisely what these Units have been trying to offer, a study of the historical dimension of science.

Had these recommendations been followed we might have expected to have a sufficient explanation for the curves in Figure 3.17, with Britain trailing behind Germany in effective scientific output. But, in fact, the opposite happened and the British universities set themselves against the Devonshire recommendations and instituted courses in science of a high degree of specialism. Unlike their German counterparts they did not appear to be unduly moved by the demand for professional chemists, for such a demand remained small for many years to come. Possibly it was the great volume of science that was now available for study that crowded out all the 'optional extras', or perhaps it was the Victorian belief in the division of labour, which decreed that concentration on one task was always better than letting one's attention wander into wider issues. Be that as it may, this kind of specialization did not produce a new generation of scientific specialists analogous to those in Germany. Basically *it was the wrong kind of specialization*. It was specialist in the sense that it concentrated upon one science, but it was not a study to the depth of the German research programmes. In a word, its shape was determined by that most unbending of all tyrants, the English examination system. Education was controlled to a large extent by what kinds of subjects were examinable in a three-hour written paper and what were not. In chemistry this meant that the traditional division into inorganic and organic chemistry, with appropriate practical tests in each, was convenient, easy to organize and not too demanding on the imaginative resources of hard-pressed examiners. And this applied from the Cambridge Tripos, through the Honours degree examinations of the new universities to the Department of Science and Art examinations for mechanics and students at the technical colleges and institutes. The consequences of this arrangement for industry are perhaps open to question, but their effects on the structure of chemistry were clear. Writing of the inorganic–organic separation of chemistry, in 1899 W. A. Tilden observed that it was

> ... unfortunate ... that the division between them, though practically necessary, should be maintained in so absolute and arbitrary a manner ... All chemists, however, now agree that there is but one chemistry so far as principles are concerned, no matter how various may be its applications. The sharp distinction and separation of inorganic chemistry and organic chemistry is in teaching and learning a source of great loss and inconvenience.

3.1.5.3 Ideology

It remains to note that sometimes the structure of chemistry has been affected by changes in the complex of opinions held by society that we sometimes call an ideology. This is often based on emotion, prejudice and other non-rational attitudes and is, of course, particularly obvious in certain fields of politics. None the less it may also have an effect on the way scientists approach their own activities. As we saw earlier, a general disenchantment with science in society, for whatever cause, can bring scientists together in a kind of defensive unity. In late Victorian Britain the public image of science went through a bad phase on account of the widespread fear that evolutionary biology would pose a serious threat to religion and morality. As it happens this does not seem to have affected chemistry nearly as much as did the earlier possibility of withheld government finance. Perhaps it was because the physical sciences in general were then so hostile to biology that their isolation from it was an adequate response to popular reaction. Certainly the structure of chemistry was little altered on that account.

In France, however, matters were different. Early in the present century there arose a strong reaction against science which was frequently asserted to be in a state of *bankruptcy*. A recent author (M. J. Nye) has spoken of 'an almost public nausea at the failure of science to cure social ills', though it had at the same time (or so it seemed) reduced man to a mere automaton and relieved him of many of the traditional comforts of religion. So strong was this French 'swing from science' that the mathematical physicist Henri Poincaré could write of the 'ephemeral nature of scientific theories' and add that the ordinary man 'concludes they are absolutely vain. This is what he calls the bankruptcy of science'.

The response of science to this crisis was by no means uniform or homogeneous. But some of the most trenchant replies came from the physicists Ernst Mach and Pierre Duhem. Both joined forces with Ostwald in identifying much of the theoretical problems in science with the mechanistic hypotheses of atoms and molecules. For Duhem, especially, science would discover its soul in thermodynamics. Ultimately they were to suffer the most crushing defeat to their anti-mechanistic idealism at the hands of Jean Perrin, who had been led by his studies of Brownian motion to assemble all the evidence he could for atomism. This, of course, is an entirely different story, but it is sufficient for us to note that the efforts of the anti-atomists were far from fruitless in that, at their hands, the divergence of physical chemistry from the more descriptive branches received an impulse that, in France at least, lasted well into the 1920s.

3.2 Reunification—PHASE VI

The Second World War inevitably had a deep effect on the progress and structure of chemistry as pursued throughout the world. As with many other aspects of life, chemistry could never be quite the same again. In fact, it was well into the 1950s before the shape of things to come became tolerably clear. Whatever stood in doubt, one aspect of chemical thinking was abundantly obvious. There were signs on all sides that the old tripartite division of chemistry was beginning to outlive its usefulness. At first the new approach was heralded by the appearance of modest publications whose first role appeared to be that of baffling the good intentions of librarians deputed to classify them as 'inorganic', 'organic' or 'physical' (or, of course, historical, analytical, technical or just general). Titles like that of C. K. Ingold's epochal work *Structure and Mechanism in Organic Chemistry* (1953) really concealed the hybrid nature of the whole enterprise described by one of its main originators. Much of the book consisted in erudite arguments based upon spectroscopy, dipole moments, thermodynamics or (above all) reaction kinetics. There were many other works, not as important as Ingold's, perhaps, but equally defying the traditional classification.

However, it was the 1960s that saw the 'literature explosion' in chemistry which bore such eloquent (and expensive) testimony to the fact that now the barricades were coming down. Just consider these new serial publications, all in English, all about a subject at the organic/inorganic interface, and all making their debut within two years:

Journal of Organometallic Chemistry	1964
Advances in Organometallic Chemistry	1964
Annual Surveys in Organometallic Chemistry	1964
Organometallic Chemistry Reviews	1965
Organometallic Syntheses	1965

It is hard to resist the conclusion, on these data, that organometallic chemistry had arrived as a new subject. Another great new hybrid was similarly celebrated in 1963 with the first volumes of *Advances in Physical Organic Chemistry* and *Progress in Physical Organic Chemistry*.

Further signs of regrouping were evident when the British *Journal of the Chemical Society* was being reorganized a few years later. Subdivision being necessary on account of the increasing numbers of published papers, it was contemplated that it might come out in three parts, corresponding to inorganic, organic and physical chemistry. Some members of the Publications Board, however, had other ideas and proposed *synthesis*, *structure* and *dynamics* as the threefold division. In the event, the Board settled for a more conservative arrangement, but dignified the organic, inorganic and physical sections by the names of Perkin, Dalton and Faraday. (The choice of the middle name was slightly eccentric since Dalton's output of what would today be recognized as genuinely inorganic chemistry was negligibly small; his one great chemical achievement—the atomic theory—spans the whole subject.) In fact, the Perkin section is now divided further into straightforward organic and physical organic.

These remarks are to remind you that there is plenty of literary evidence for the state of flux in which chemistry now finds itself, even as regards its own internal organization. But the evidence does not stop there. It may also be found in the approaches to the presentation and teaching of chemistry that were mentioned in passing in Unit 1, Section 1.0.4. Numerous courses have appeared, not only in our own university, in which traditional divisions have been questioned or even abandoned. And, in Britain and elsewhere, several more Chairs have been created in the last few years in which the occupants are not required or expected to identify themselves with any one of the traditional branches of chemistry.

We cannot expect to find that such radical regroupings take place overnight, and there is plenty of opposition to the abandonment of what is often seen as a well-tried approach to the subject. What we are arguing is not that chemistry has suddenly become one subject (whatever that might mean) but that the organic and inorganic branches are now converging; when, if ever, they unite is in the unpredictable future. But after such a long period of divergence and isolation it becomes urgently necessary to ask what has given rise to the changes we see.

Figure 3.20 Sir Christopher K. Ingold (1893–1970), English chemist and one of the founders of the modern electronic theory of organic chemistry. From 1923 to 1930 he was Professor of Organic Chemistry at Leeds, and from 1930 until 1961 he was at University College, London. Here he developed the distinction between S_N1 and S_N2 reactions, pioneered studies in aromatic electrophilic substitution, established the complexity of elimination mechanisms and explored in great depth a multiplicity of organic rearrangements. With characteristic emphasis on the unity of chemistry, Ingold inaugurated the tradition of a single series of research colloquia at University College.

It is the contention of this Unit that there have been three fundamental factors at work. In the following discussion of these you may note several terms that are not wholly familiar at this stage. Fuller explanations will be found later in the Course.

3.2.1 The rise of organometallic chemistry

For all practical purposes organometallic chemistry can be said to have originated with Frankland who, among other achievements, invented the term (see Unit 2, Section 2.1.2). For a few hectic weeks the zinc alkyls that he discovered played a critical role in the emergence of one of chemistry's cardinal doctrines, the theory of valency. Thereafter they played a modest part in the development of organic synthesis, until the discoveries of Grignard sent them to near-oblivion. For another half century the organomagnesium compounds were to prove one of the most versatile types of reagent ever available to the organic chemist. Then, almost exactly a century after Frankland sent his paper to the Royal Society, came one of those rare events that historians can with a good conscience hail as a 'breakthrough'. The odd thing is, though, that few people at the time were conscious that a breakthrough was much needed. The event in question was the discovery of ferrocene.

In 1951 two groups of workers, independently but almost simultaneously, discovered a highly stable orange solid, $Fe(C_5H_5)_2$, capable of being vaporized unchanged and yet containing only carbon, hydrogen and iron. It was prepared from cyclopentadiene either by the action of a Grignard reagent on ferric chloride or by just heating with finely divided iron to 300 °C in nitrogen. Christened at the time 'ferrocene' it has since been dignified by the systematic name bis(*pentahapto*cyclopentadienyl) iron. Apart from being the first reported compound to contain only carbon, hydrogen and iron, it was remarkable for its stability, its analogous reactivity to that of aromatic hydrocarbons and (above all) its structure. At first this was written as

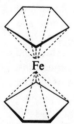

which is an entirely conventional arrangement. The shattering revelation that this, like all other classical formulations, was wrong came with the X-ray analysis of the crystal, which revealed the iron atom sandwiched between the two parallel cyclopentadiene rings:

Lewis' bonding theory could certainly not explain that.

In a note on this discovery and that of the ruthenium analogue in the following year's *Annual Reports* (inorganic section), the Reporter announced 'the discovery of these compounds opens up a new field in the borderland between inorganic and organic chemistry'. History has vindicated his judgement, but we may well ask why he could have been so sure. In seeking for an answer we shall need to note a characteristic feature of chemistry that appears from time to time.

Before 1951 organometallic compounds had posed a large number of difficult questions. Grignard reagents, apparently the simple $R-Mg-X$, were obviously much more complex systems of uncertain nature. Various studies had come up with what seem like isolated cases of compounds between transition metals and organic molecules (including the lone example of a combination of ethylene and platinum chloride reported by Zeise as far back as 1831). During the 1920s numerous mysterious compounds had been isolated by Hein, apparently involving chromium and phenyl radicals. From the 1930s, various solvent phenomena had suggested some kind of interaction between the organic solvent and a metallic compound, as when solutions of aluminium halides in alkyl halides were found to have quite high electrical conductivities. Most familiar of all were the compounds formed between transition metals or their compounds with carbon monoxide, which, after all, was normally banished from the organic literature by a very arbitrary convention.

It seems that there was an accumulation of unrelated facts all looking for a theory to explain them. Such a theory was actually to hand—the molecular orbital theory—but before it could, or rather would, be applied a strong new incentive was needed. This was provided in full measure by ferrocene, whose spectacular properties were the wonder of the day. Given this impetus a generalized molecular orbital treatment of organometallic compounds could be developed, the earlier puzzling cases could either find their first clear explanation or could be reinterpreted in a more subtle way, and a host of new experiments could be suggested. This, in fact, is just what did happen, the ligand field treatment being based on molecular orbital methods combined with an earlier crystal-field theory developed before the last war for magneto-physics by Bethe and others (see Unit 14).

Since that time organometallic research has expanded to an extent indicated by the literature explosion of the mid-1960s. A glance through Cotton and Wilkinson will give some impression of its importance as viewed from inorganic chemistry. Few metals have not been the subject of enquiries into their possible links with organic residues. But the most profound effects have probably been in its impact upon the structure of chemistry itself. The following quotation (from a standard textbook, *Organometallic Chemistry*, edited by H. Zeiss) must seem remarkable for its early date, 1960:

> The extraordinary convergence of organic and inorganic chemistry in the metallocene and metallarene structures has provided a new area of mutual interest shared by experimental and theoretical chemists... The universality of effort and the increase in scientific intercourse between inorganic and organic chemists permit the optimistic view that the artificial barriers which existed formerly between organic and physical disciplines are being leveled at this border also.

> SAQ 14 (*Objective 10 and revision*) Both the zinc alkyls and ferrocene may be said to have contributed to the unification of chemistry in the sense that both could be designated as organic or inorganic. What other aspect of their chemistry helped to bring a new unity to the subject?

If a personal judgement may be permitted in relation to organometallic chemistry at the present time it would be this: so remarkable has been its growth that it really does make sense to regard it as a new branch of chemistry in its own right. Its success has assured a considerable degree of mutual interpenetration by the organic and inorganic branches. But to assume that for this reason the two are now fully integrated is to ignore many areas of research when the insights of one are largely irrelevant to the other. In due course you should be able to identify some of these; meanwhile it is advisable to treat claims of complete integration by organometallic chemistry with some scepticism.

3.2.2 The overspill of physical chemistry

By this phrase we mean the failure of physical chemistry to be contained any more within a rigid framework and its employment as a routine tool in the exploitation of those areas of chemistry traditionally labelled 'organic' and 'inorganic'. We shall argue that in changing its own role it has also altered in a radical fashion the nature and the interrelationships of the other two branches.

Now it will be recalled that one effect in physical chemistry in its early days was to syphon off those topics from the other two areas that might have contributed to their union. This was undoubtedly true when the 'syphoned-off' parts were relatively small. Once physical chemistry had permeated the majority of the organic and inorganic areas, however, its effect would be to demonstrate that they had one more thing in common—an accessibility to the techniques and interpretations that we now label 'physical'. Whether this is plausible may be debated, but we shall only have time to produce a small amount of the evidence available.

Organic chemistry was the first to display widely the impress of physical theory. As we have seen, many of the organic chemists between the wars were reluctant to use physical methods of analysis in solving their own constitutional problems, although measurements of optical refraction and dipole moments came slowly to be accepted. The theoretical treatments associated with quantum theory and wave mechanics became part of the organic chemist's stock-in-trade only after the last war, though Sidgwick, Pauling and others in the prewar period had done valuable work. The use of reaction kinetics to study organic mechanisms owes much to Ingold, who while Professor of Organic Chemistry at Leeds from 1924 and later at University College, London, from 1930, was able to build on the foundations laid by Robinson and bring in many insights from kinetics to yield a new and comprehensive theory of organic mechanisms, especially in the area of aromatic nitration, ester formation and hydrolysis, alkyl halide replacement and molecular rearrangements. His book *Structure and Mechanism in Organic Chemistry* was designed for university undergraduates and profoundly influenced the teaching of organic chemistry after its publication in 1953.

Identification of physical organic chemistry as a topic in its own right began with a book of that title by Hammett in 1940, urging that 'a unified and consistent treatment is possible in terms of a few simple generalisations and theories'. The appearance in 1963 of the *Progress* and *Advances* series referred to earlier owed much to Hammett's initiative during the war years and to Ingold's teaching and research well into the 1960s.

Nor was organic chemistry the sole beneficiary of physical theory. During the 1950s the oldest branch of the subject underwent a change so radical that one of its distinguished practitioners, R. S. Nyholm, announced 'the renaissance of inorganic chemistry'. In his inaugural lecture (1956) for the Chair at University College, London, he identified two causes for the 'rebirth' of the subject. One was the application of quantum mechanics and the other the extensive use of physical methods for structure determination. In the broadest sense these were both extensions of physical theory to inorganic chemistry.*

As with organometallic chemistry, the new insights could be applied to many outstanding problems of the past. There was another case of a 120-year gap (similar to that for Zeise's salt) between discovery and recognition, this time for the first example of quadruple metal–metal bonding, identified in 1964 for a compound reported in 1844. The reinterpretation of Werner's work on coordination compounds brought such complexes to the fore and established them as one of the central themes of inorganic chemistry.

This brings us to a further aspect of the subject, that many of the compounds and reactions now being studied have much more in common with those familiar to the organic chemists than with the salts, oxides, sulphides, etc. of traditional inorganic chemistry. Coordination and organometallic chemistry have combined to reveal the great importance of the covalent bond for the inorganic chemist. The consequences are obvious. Despite—even because of—its renaissance as a subject, inorganic chemistry has now so much in common with the organic branch that for many people it makes little sense to keep the two in their traditional isolation. In finding its identity inorganic chemistry may seem to have lost it, though perhaps it would be truer to life to put it the other way round. Nyholm has an interesting sentence: 'The modern inorganic chemist has scant regard for the distinction between inorganic and physical chemistry'. But he still talks of an *inorganic chemist*! Similarly, when T. Moeller introduced his *Inorganic Chemistry* four years earlier (in 1952), he accepted that in the last thirty years 'lines of demarcation with other phases of chemistry have become less well defined', but he was very keen to emphasise that 'inorganic chemistry is not general chemistry'.

One last point that must be made regarding this overspill of physical chemistry concerns the role of instrumentation. The early neglect of spectroscopic data, with useful information for both organic and inorganic chemistry, was due to the sheer tedium of collecting and interpreting them. Anyone who has plotted by hand an infrared or ultraviolet spectrum will begin to understand the problem.

During the 1920s and 1930s this was certainly true of visible and ultraviolet spectra. Even the physical chemists were deterred by problems of detailed interpretation. Today one of the values of uv spectra is their analogical use.

* A third reason for this 'renaissance' has been seen in the growth of atomic research following the 'Manhattan Project' of the Second World War: this has created considerable demand for inorganic chemists, as well as those more accurately styled 'radio-chemists'.

Figure 3.21 Sir Ronald Nyholm (1917–1971), Australian chemist. In 1951, Nyholm became head of the chemistry department at University College, London, his inaugural lecture *The Renaissance of Inorganic Chemistry* being an accurate indicator of both his interest and his hope. In this area of chemistry he had worked both in Australia and in England, being specially concerned with the application of physical techniques to inorganic problems. He was closely involved with the Foundation Science Teaching Project.

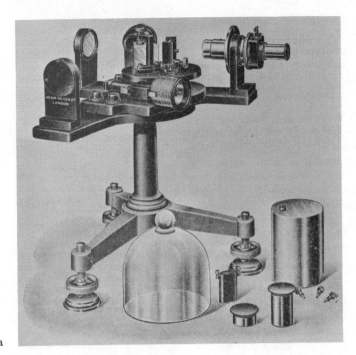

a

b

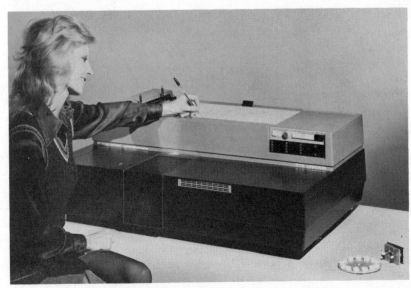

c

Figure 3.22 The development of the infrared spectroscopy. The first instrument (a) is by Hilger, 1928. It has a rock-salt prism and is of course a single-beam non-recording manual spectrometer. The second instrument (b), also by Hilger, appears to be from the 1930s and has the great advantage of being a recording model—one of the earliest examples of such an apparatus. Both of these instruments contrast vividly with the third instrument (c), one of the latest infrared spectrophotometers, the Pye Unicam SP2000. Its interlinked push-button controls, with automatic chart synchronization, automatic superimposed and consequential scanning and its extensive computer control facilities combine to ensure 'foolproof operation'. The use of caesium iodide cell windows enables information to be obtained down to 200 cm^{-1}, of great value to the inorganic chemist.

Without the necessity for detailed interpretation it is often sufficient to compare a good spectrum with a whole series of spectra from known substances. But this can be useful only when large numbers of such spectra are available, and this in turn depends upon ease of production. It was not until after the Second World War that the use of photoelectric recording instruments became available for uv spectroscopy and this offers further insight into the significance of the 1950s for the structure of chemistry.

An even more extreme case is that of optical rotary dispersion. Many earlier polarimetric studies of optically active compounds were performed with light of constant wavelength, often the D-line of sodium. But by 1896 Cotton and others were carrying out quite detailed studies on the way in which rotation varies with wavelength. The shapes of the resultant curves were shown to give valuable information about the steric configuration of those parts of the molecules that were absorbing radiation. However, it was not until the 1960s that the technique became common, the enormous difficulties of obtaining such spectra having been largely overcome by the use of recording uv spectropolarimeters. The related phenomenon of circular dichroism has recently acquired an importance even greater than that of optical rotatory dispersion.

Two other instrumental techniques that have played an important integrating role in recent years are nuclear magnetic resonance (nmr) (see Units 7 and 8) and mass spectrometry. The wide extension of nmr techniques reinforced the effect of uv and ir spectroscopy in providing insight into structural problems of both branches. Mass spectrometry was a rather different case in that for several decades it was applied almost exclusively to inorganic compounds, but in the last twenty years has been increasingly valuable in determining organic structures, for which it has developed its own specialist methodology. Nevertheless, like nmr, this technique has materially helped to draw together the two branches.

Examples of this kind can be multiplied almost endlessly. It is clear that instrumentation has played a decisive part in reunifying chemistry. Why this should have happened in the 1950s and 1960s is a complex matter, but three things seem to be important. One is the legacy of research in electronics inherited from development of wartime radar; another was the invention of the transistor in the late 1940s, with all the attendant developments of microcircuitry; the third is the growing recognition that science needs money and that expensive equipment is not necessarily a luxury. On that last note of sober realism we conclude this Section and turn finally to consider briefly some of the external influences of the current convergence within chemistry.

SAQ 15 (*Objectives 1 and 9*) The graph (Figure 3.23) relates to the growth of physical organic chemistry reflected by the number of items on that subject in *Chemical Abstracts*. What general inferences can be made?

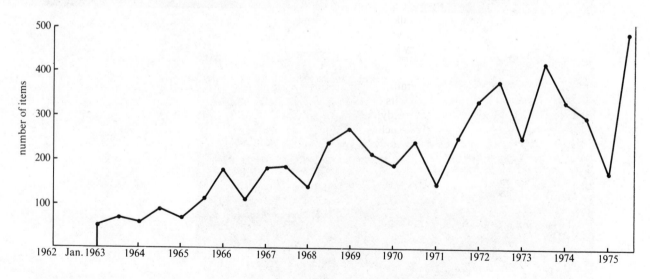

Figure 3.23 The growth of physical organic chemistry. This graph charts the rise of this subject in the last few years by plotting against date the number of items so classified in *Chemical Abstracts*.

3.2.3   Some social factors

In the changing post-war world two external factors have strikingly affected the structure of chemistry, and both have acted in the same direction.

The first of these is money. After 1945, science could never remain—if it ever had been—an isolated thing. Massive investment in science became an item of government policy and private benefaction in many countries in the West. The number of scientists was far less than required and prodigious sums were spent in establishing new institutions and laboratories. One consequence was that more money was available than ever before for the purchase of scientific equipment, and this in turn led to the development of increasingly sophisticated instruments. As we have noted above, the growth of instrumentation was a major factor in the convergence within chemistry. To that extent one must acknowledge as an important factor the far greater availability of money.

Now, of course, those opulent days are something of a chemist's nostalgic dream. In any case they probably never were quite as good as he remembered them. But they were good enough by comparison with the frugality of the 1970s. And this brings us to a second factor, very much of our own time.

No one can deny that in society today there are strong currents of opinion inimical to science. Their causes are multifarious. In some respects the situation in the mid-1970s is comparable with that in France at the turn of the century. In others it is more like that of the Middle Ages from which modern science emerged at the Renaissance and Reformation. But its effects are inescapable.

In the first place there are demands that chemistry, in particular, should pay far more attention to the effects of its products on the environment. So far as ecological chemistry may be said to exist it does so without any reference to traditional categories of chemical structure. After all one can be poisoned just as efficiently by inorganic arsenic as by organic systemic insecticides. Equally, atmospheric pollution owes at least as much to the carbon monoxide from the combusted organic fuels as it does to the presence of sulphur in lead compounds derived from their inorganic additives. And in the world energy crisis men look with equal anxiety to the discovery and exploitation of more organic fossil fuels and to the possibilities of nuclear energy dependent upon compounds that are inorganic as well as fissile.

Nor is it only ecology that challenges the chemist. He is, and should be, very concerned about the anti-scientific, anti-intellectual, anti-rational streak in much popular thinking which tends to regard science as frankly a bad thing; the least of reasons why he should be anxious is that it may well affect his job. This is no isolated phenomenon in Britain, and one of its global effects has been to unite scientists generally in the defence of their values. It would be absurd to suggest that science was in a state of 'siege economy', but it is under widespread pressures, and there is nothing like this kind of situation to bring people together in common cause. Hence we find much talk in schools and elsewhere about 'integrated science', though often, one feels, with an optimistic hope that a different blend of physics, chemistry and biology is all that is needed to set things right again.

Chemistry has been particularly vulnerable, partly because of its obvious relevance to ecology but partly also because its privileged status in the hierarchy of sciences looks in some danger. It is already seen by some as a kind of 'handmaiden' to rapidly growing new sciences such as biochemistry, geochemistry and the like. The result is that chemists are now displaying a most commendable sense of fraternal unity, and many articulate spokesmen are proclaiming that chemistry is now one in a sense that was never true before.

The extent to which these factors have been important in favouring convergence within chemistry is, of course, a matter for one's own value-judgement, partly because they cannot be measured in any obvious way and partly because we are still so close to the events concerned. But it would seem foolish not to remind ourselves that even events in the world of ideas apparently remote from chemistry can affect the structure of chemistry itself.

3.3 Conclusion

To conclude this Unit, and *The Structure of Chemistry*, we offer a few thoughts on what is intended by a phrase that has recurred so frequently that it is possible that we have overlooked its varied shades of meaning: *the unity of chemistry*. As we have seen, there have been six phases of development of our subject where this 'unity' has been alternately dominant or recessive. Largely, this has been a question of how far organic and inorganic chemistries have been converging or diverging. In the one case, union looks likely, in the other the odds seem stacked against it. But what exactly is the unity of chemistry? The simple word *unity* itself can be used in so many different contexts that it has acquired all kinds of overtones, not all of which are helpful when considering the unity of chemistry. In its applications to *people* it is usually employed in a rhetorical sense, with the implication, as in politics, church affairs or team games, that it is always a good thing. It by no means follows that when applied to chemistry this should *necessarily* be the case, though it may well be so. But it needs to be demonstrated.

Now it so happens that one can often discover what people mean by an assertion by asking what it is that they are trying to deny. When, therefore, chemists have announced that 'chemistry is one' it will do no harm to find out what propositions they are anxious to contradict. If we do so we find an interesting range of opinions.

In the first place we sometimes discover that they are really saying that chemistry should not be treated as a collection of unimportant bits and pieces but as a serious undertaking worthy of patronage and support. Logically, of course, there is no reason to suppose that the whole is greater than the sum of its parts, but financial supporters are not always most effectively wooed by mere logic. It is not entirely cynical to suppose that these considerations played some part in the struggles for government patronage during the past 150 years. In so far as they tell us anything about their authors' views of chemistry it is only on the level of its *organization*. Clearly a centrally organized group of laboratories is more attractive to an investor, public or private, than a motley collection run by entrepreneurs in isolation from one another. So we must beware lest we are beguiled by an ecumenical battle-cry into supposing that the issue really concerns the *nature* of chemistry.

However, the proponents of a united chemistry usually mean more than that. They may be trying to deny the notion that *the techniques and methods* of the two main branches are so different from each other that there is little chance of expertise in one being of value to the other. This kind of objection will, of course, increase in force as time goes on and specialist techniques get more complex. From about the 1840s the technical differences between organic and inorganic manipulations have continued to grow and, were it not for mitigating factors on the other side, the advocates of 'one chemistry' would have had an increasingly difficult task in making themselves heard. In fact, by now they would have been shouting their heads off. We shall remind ourselves of those factors in a moment.

To assert 'there is but one chemistry' may be to protest at the notion that the *basic theoretical principles* of organic and inorganic chemistry are not the same. This is a third way in which the phrase can be used. We may therefore reasonably enquire what basic principles of this kind have been at stake? In organic chemistry there was at one stage the possibility that vitalism reigned, but gradually this barrier to unity collapsed. Later on, each branch developed plenty of unique theoretical principles: the catenation of carbon atoms and the periodic law are just two. But, looking back, it seems that the apparent possession of such principles by one branch and not the other has never been a major obstacle to attempting a comprehensive view of chemistry as a whole. However, the very existence of such a thing as 'organic chemistry' implies at least some unique features; otherwise the question would not be one of unity but of identity.

We may thus conclude that the absence of universal principles has rarely been a major barrier to a recognition of the essential theoretical oneness of chemistry. It is rather that where the *presence* of common features *is* shown the sense of unity returns. This was exactly the case in phases II, IV and VI. But it seems that this has to be reasserted from time to time, either because this sense has worn thin over the years or because some of the supposed features of unity have been subsequently shown to be spurious.

When the unity of all chemistry is asserted, therefore, it is this last sense that is the most fundamental: that chemistry has running through all its branches certain common principles. Generally speaking these principles will not be the same at all times. We have seen what happened to Berzelius' electrochemical theory. Even atomism failed to cement the subject together in the last few years of the 19th century, and chemistry existed long before Dalton in any case. Nor can we quote energetics, for these were largely absent in the early years of our period, and in any case they cannot differentiate chemistry from physics. But there does seem to be one recurring motif which has never been entirely absent but which seems to give to chemistry that unique quality that differentiates it from all other sciences, and yet which is common to all its branches. It is the concept of *chemical individuality*, and its derivative, that of *chemical purity*. Perhaps it has been the unconscious recognition of this most distinctive facet of chemical thought that has, all along, inspired the optimistic belief that, in these basic terms, chemistry really is one.

And what of the future? If the model of alternate convergence and divergence is valid we should be in for a period of divergence, but no time-scale can be laid down. However, there are signs that the present era of integration of organic and inorganic chemistry may not go on for ever. If the swing from science is halted, as many in Britain think it will be, one strong external constraint on chemists is removed. As we have seen, there is nothing like opposition from outside to unite those within. Those experiments on advanced chemical education where chemistry is taught as an integrated whole have not all been an unqualified success, though whether this is because of built-in conservatism or of some inherent defects it is not yet possible to say. We hope that when you have finished this Course you will have your own views of what the future structure of chemistry may be. Perhaps you will conclude that the day may come when chemistry will once again be recognized as a great tripartite assemblage of *structure*, *synthesis* and *dynamics*.

Or it could be *inorganic*, *organic* and *physical*.

SAQ answers and comments

SAQ 1 This ignores:

(a) the prior existence of journals which concentrated on one branch of chemistry without indicating as much in their titles; *Berichte* is a classical case;

(b) the danger of generalizing from the output of one country only—all of these titles originated in the USA;

(c) the possible variations in editorial policy of existing journals, economic circumstances, etc. See the further discussion in the text.

SAQ 2 The advantages were believed to be the preservation of constant valency for:

(a) and (b) nitrogen, 5-valent as it was thought to be in NH_4Cl.

(c) mercury, 2-valent as in $HgCl_2$

(d) sulphur, 2-valent as in H_2S

(e) platinum, 4-valent as in $PtCl_4$

An argument against these could have been that the stratagems caused more problems than they solved, e.g. 1-valent oxygen in (b) and 2-valent chlorine in (e). But the real evidence came from vapour densities and chemical reactions. Vapour densities implied monomeric formulae for ammonia, nitric oxide and mercurous chloride, though in the latter case a solution was found by Odling's discovery that gold-leaf held in the vapour was amalgamated, suggesting a dissociation on heating, i.e.

$$Hg_2Cl_2 \rightleftharpoons HgCl_2 + Hg$$

rather than
$$Hg_2Cl_2 \rightleftharpoons 2HgCl$$

Chemical evidence began to multiply against such formulae. Thus (d) was rejected for benzenesulphonic acid by the fact that this compound is reduced to thiophenol rather than phenol, suggesting the phenyl group is linked directly to sulphur (Vogt, 1861):

$$C_6H_5SH \quad \text{thiophenol}$$
$$C_6H_5OH \quad \text{phenol}$$

SAQ 3 There was the general problem as to what exactly was meant by 'latent atomicity'. Was there any merit in thus labelling something which was by its very nature undetectable? The discoveries of new compounds of V, W, Ta, Nb and other transition metals revealed that the valencies of the metals often varied by *one* at a time (e.g. VCl_2, VCl_3, VCl_4, V_2O_5) not *two* as Frankland's theory would require. Frankland's examples were fortunate in that iron(II) chloride was shown to be monomeric while iron(III) chloride was dimeric in the vapour phase.

SAQ 4 The pair of reactions carried out in 1876 by Victor Meyer and Lecco (see Unit 2, Section 2.4.4).

SAQ 5 The notion of molecular compounds, with two components loosely held together, is in direct line with Berzelius' electrochemical theory (cf. the formulation of salts, Unit 1, Section 1.2.1.4). In so far as this idea has survived in the modern doctrine of association by intramolecular hydrogen bonds, as in $(HF)_n$, our own views are completely Berzelian. But the notion of constant valency springs much more from the theory of types. As Schorlemmer remarked in 1879, 'It is easy to see that the theory of types must have led to the conception of a constant atom-fixing power'.

SAQ 6

(a) Overall valency $= 3 - 0 = 3$

(b) Overall valency $= 3 - 3 = 0$

(c) Overall valency $= 4 - 3 = 1$

SAQ 7 By asserting that carbon happens to have its valency equal to its coordination number.

SAQ 8 (a) This implies that some barrier, hitherto unbreached, had at last crumbled away, presumably the one which confined optical activity to the organic side; now it was shown to be a phenomenon existing in *both* fields. But that was relatively a trivial matter, not to be compared with the large issue of vitalism in the previous century. A more fundamental—though less obvious—point is that the successful optical resolution could be seen as a vindication of Werner's stereochemistry *and therefore of his coordination theory in general*. This, as we have seen, would reduce the chemistry of carbon compounds to the status of being a special case of coordination theory. (b) However, we have met many claims already of 'completion' of the unification initiated by Wöhler, and the characterization of that notable research in terms of 'demolition' is a highly suspect operation. Finally, as we shall go on to show, this was not at all the effect on chemical thinking or organization that actually happened. Inorganic chemistry was by no means to be confined to the coordination complexes which, by their very nature, were long regarded as rather exceptional. This kind of exaggerated statement can result from a sincere desire to render credit where it is due, but also from an incorrigible tendency many of us have of applying to the past modern chemical ideas and attitudes. Because *we* are aware of the widespread importance of coordination chemistry it does not automatically follow that this was the case sixty or more years ago. In fact it was not.

SAQ 9 Nyholm is really making the same point as in TV programme 1, that progress in science does depend upon the availability of proper techniques. The ones intended are presumably those of instrumental analysis (which could lead to structural conclusions). Others, involving the boron hydrides, were those of high vacuum technology, of value in the preparation and isolation of volatile and unstable individuals, some of which had great theoretical interest.

SAQ 10 By coming to the aid of natural product investigations it gave these a powerful new boost and at the same time relieved organic chemists of the necessity of relating their work to that in the inorganic field. In so far as organic chemistry did become preoccupied with natural products it tended to look more to biology than to the chemistry of the 90 odd other elements. Divergence rather than convergence was the mark of this application of synthesis.

SAQ 11 In 1831 organic chemistry was about to be compared to the 'primeval forest' of Wöhler (Unit 1, Section 1.2.3.1). The year 1871 is a more promising date in that the theory of structure was only 10 years of age and the cyclic formula for benzene not much older. On the other hand the problems of the fine structure of benzene, of unsaturation and of tautomerism were still to be resolved, while the tetrahedral hypothesis of van't Hoff, destined to inject so much order into the chaos of isomerism, was still three years away. By 1901 organic chemistry had certainly forfeited the right to such confident affirmations as these. It was distinctly light on the theory, and no new spectacular advances on the theoretical side had taken place for a long time. We are therefore left—correctly—with 1931 and the new electronic theories as the substance behind the satisfaction so complacently expressed by A. W. Stewart.

SAQ 12 Obviously any answer will be a value-judgement, but certain elements ought to be fairly clear.

Although we have deliberately refrained from going into detail it is clear that Nernst's paper contained a good deal of closely-reasoned mathematics. As we have seen, this would present severe difficulties to many chemists from the organic or inorganic branches. Secondly, the paper makes no claims to finality and almost invites the application of further data. Specific heats at low temperatures were hardly the stock-in-trade of most traditional chemists, and they would not feel under much compulsion to hunt for these. Thirdly, although Nernst (rightly) asserted that his theorem offered important connections between the data of thermochemistry and the all-pervasive problem of chemical affinity, it did so in terms of the wide generality characteristic of the physical chemists. How far these generalizations applied to the specific problems of the organic chemists was, for many of them, a moot point.

If, therefore, you concluded that the immediate response to Nernst's heat theorem *outside physical chemistry* was fairly muted, you would be largely correct. Such diverse processes as the interconversion of sulphur allotropes, hydration of salts, the combination of nitrogen and hydrogen, and the 'cracking' of petroleum fractions were all studied in the light of Nernst's work within a fairly short time. Yet in almost every case the work was done by those who thought of themselves as physical chemists rather than anything else. Some work was of great industrial importance (especially the last two examples above) and affected Germany's wartime prowess. Much was due to the efforts of Fritz Haber, Professor of Physical Chemistry at Karlsruhe from 1906 to 1911 and then Director of the new Kaiser Wilhelm Institut in Berlin, devoted to the study of physical chemistry and electrochemistry.

SAQ 13 (a) In British manufacture of sulphuric acid before 1900. There was an enormous increase in production which even surpassed that of Germany. Presumably this reflects two facts:

(i) Sulphuric acid could be produced by rule-of-thumb methods without the necessity of R & D programmes.

(ii) The British educational system, in comparison with Germany, was not in any case geared to industrial needs to any degree at the time.

One may also note the sharp decline in sulphuric acid production in both countries during the 1920s, reflecting the worldwide economic depression of those years. Education responded much less dramatically.

(b) The changes in dyestuff production, especially in Germany, responded far more violently to the upheavals of two wars and the inter-war depression. Here we see the importance of patent restrictions and capital investment.

(c) The one prediction that still seems valid is that the economic prosperity of the chemical industry, and therefore of the countries concerned, is going to depend on adequate training and education.

SAQ 14 Both zinc alkyls and ferrocene focused new attention on the problem of *bonding*. In the former case, Frankland was led to his conception of valency and to the designation of a *chemical bond*. With ferrocene, more than one hundred years later, the new type of combination between the iron atom and the cyclopentadiene rings led to a new understanding of links between π-bonding systems and transition metal atoms and to the application of molecular orbital theory to inorganic and organometallic compounds on a far wider scale than before.

SAQ 15 First, there is the obvious inference that physical organic chemistry has manifested a steady overall growth from 1963 to 1975. Secondly, there are, of course, peaks and troughs, but no trough as deep as the one in 1975, which undoubtedly reflects the economic situation in the western world. The remarkable recovery for the second half of 1975 indicates that during the previous few months underemployed chemists have been spending time writing papers about their past research. Then, again, there is the fact that, with this indicator, physical organic chemistry will appear to have been born in 1963. As we know this is not the case, and the long gap between the first publication of Hammett's *Physical Organic Chemistry* in 1940 and this date indicates the hazards of using one kind of indicator only. In fact, it would seem that 1963 marked the beginning of the recognition of this subject as a distinct specialism, a fact entirely consistent with the appearance of the *Progress* and *Advances* series in that year.

Further reading

The following suggestions are again for those who wish to read further in the subject: the first four are general works which you have met before, and the remainder are more specialist treatments of themes within this Unit.

A. Findlay and T. I. Williams (1965) *A Hundred Years of Chemistry*, Methuen, London, chapters 5–13.

A. J. Idhe (1966) *The Development of Modern Chemistry*, Harper and Row, New York, Evanston and London (frequent references).

J. R. Partington (1964) *A History of Chemistry*, vol. 4, Macmillan, London (frequent references).

John Read (1957) *Through Alchemy to Chemistry*, Bell, London, chapter 10.

R. G. A. Dolby (1968) The emergence of a speciality, a case study: physical chemistry, *Actes du XIIe Congrès Internationale d'Histoire des Sciences*, Paris, **vi**, 29–32.

G. B. Kauffman (1974) Alfred Werner's research on optically active coordination compounds, *Coordination Chemistry Reviews*, **12**, 102–49.

G. B. Kauffman (1975) The first resolution of a coordination compound, *van't Hoff–le Bel Centennial, Am. Chem. Soc. Symp. Series No. 12*, 126–42.

R. S. Nyholm (1956) *The Renaissance of Inorganic Chemistry*, H. K. Lewis, London.

W. E. Palmer (1965) *A History of the Concept of Valency to 1930*, Cambridge University Press, chapters 6–8.

C. A. Russell (1971) *The History of Valency*, Leicester University Press, chapters 10–17.

Acknowledgements

Grateful acknowledgement is made to the following sources for material in this Unit:

Figures 3.1, 3.8, 3.10, 3.14 and 3.15 Chemical Society Library Portrait Collection; *Figure 3.3* University of Manchester Institute of Science and Technology; *Figure 3.6 Les Prix Nobel 1913*; *Figure 3.7* Courtesy of Dr George B. Kaufmann, California State University, and Victor R. King; *Figure 3.9 Les Prix Nobel 1912*; *Figure 3.11* Lafayette Ltd, from A. Findlay and W. H. Mills (eds), *British Chemists*, The Chemical Society, 1947; *Figure 3.12* H. A. Bumstead and R. G. Van Name (eds), *The Scientific Papers of J. Willard Gibbs*, vol. 1, Longmans, 1906; *Figure 3.13* F. J. Moore, *A History of Chemistry*, McGraw-Hill, 1939; *Figure 3.16* Ernst Bäumler, *A Century of Chemistry*, Econ Verlag, Dusseldorf, 1968; *Figures 3.17, 3.18 and 3.19* From G. W. Roderick, *The Emergence of Scientific Society in England (1800–1965)* Macmillan, London and Basingstoke, 1967 edn; *Figures 3.20 and 3.21* Godfrey Argent, London; *Figure 3.22a* From a Hilger catalogue of 1928 by courtesy of Professor W. C. Price, King's College, London; *Figure 3.22b* Rank Hilger, Margate, Kent; *Figure 3.22c* Pye Unicam Ltd, Cambridge; *Figure 3.23* L. R. A. Melton.

THE NATURE OF CHEMISTRY

THE STRUCTURE OF CHEMISTRY

1 One chemistry—or two?

2 Synthesis and structure

3 Specialism and its hazards

CHARACTERISTICS OF ORGANIC CHEMISTRY

4 Characteristics of carbon compounds

5 Orbital and electronic theories of organic chemistry

PHYSICAL METHODS AND MOLECULAR STRUCTURE 1

6 Vibrational and electronic spectroscopy

7 Nuclear magnetic resonance spectroscopy 1

8 Nuclear magnetic resonance spectroscopy 2

METHODOLOGY OF ORGANIC CHEMISTRY

9 Carbonyl compounds 1

10 Carbonyl compounds 2

11 The chemistry of the amino group

TRANSITION METAL CHEMISTRY 1

12 Introductory chemistry

13 Structure and stereochemistry of metal complexes

14 Bonding in metal complexes

TRANSITION METAL CHEMISTRY 2

15 Thermodynamics

16 Inorganic reaction mechanisms